The Chittagong Hill Tracts
Man-Nature Nexus Torn

The Chittagong Hill Tracts

Man-Nature
Nexus Torn

Editor
Philip Gain

Language Editing Assistance
Brother Jarlath D'Souza

Editorial Assistance
Ainud Sony and Sabrina Miti Gain

Contributors

Prof. Haroun Er Rashid, Jenneke Arens, Devasish Roy, Raja Tridiv Roy (late), Philip Gain, Sudibya Kanti Khisa, Sarder Nasir Uddin, M. Monirul H. Khan, Ronald R. Halder, Sayam U. Chowdhury, Buddhajyoti Chakma, Partha Shankar Saha, Prof. Sadeka Halim, Asfara Ahmed, Han Han, AKM Muajjam Hossain Russel, Tahmid Huq Easher, Shekhar Kanti Ray, Md. Safiullah Safi, Syeda Nusrat Haque, Supryio Chakma, Alimul Haque, Robert Alec Lindeman, Asif Khan, Tania Sultana, Ushing Prue, Nimaprue Marma, Ching Mo Sang, Lucky Chakma, and Ainud Sony.

Published by
Society for Environment and Human Development (SEHD)
1/1 Pallabi (5th Floor), Mirpur, Dhaka-1216, Bangladesh
T: 880-2-8060636 F: 880-2-8060828
E: sehd@sehd.org Website: www.sehd.org

Published: 2013

Cover photos: Philip Gain
Cover design: Goutam Chakraborty
ISBN: 978-984-8952-04-7

Printed by: KARUKRIT
Compose and page layout assistance: Lucky Ruga and Prosad Sarker
Maps: Raquib Ahmed and Dilara Hasan

The Society for Environment and Human Development (SEHD), a non-profit Bangladeshi organization, was founded in 1993 to promote investigative reporting, engage in action-oriented research, assist people think and speak out. *The Chittagong Hill Tracts: Man-Nature Nexus Torn* is a report on the state of the environment in the CHT. The reports, analyses, critiques, maps, and images published in this report portray the unprecedent ecological disaster that has ravaged the majestic beauty of the hill region of Bangladesh and severely upset the indigenous peoples who have lived there for generations.

Price: TK.500.00 US$20

Contents

Preface

The Chittagong Hill Tracts (CHT) has lost its majestic look. Once part of a mega-diversity zone in South Asia, the region is now faced with an ecological disaster. The hills are barren today. The forest resources have decreased to such great extent that the official logging that Bangladesh Forest industry Development Corporation (BFIDC) performed for decades has come to an end.

The Kaptai Dam, Karnaphuli Paper Mill (KPM) and other 'development icons' manifest concrete evidences of the ecological devastation today. The indigenous peoples who had a free run in the forest for generations now witness their nexus with nature torn. This is an unprecedented ecocide for the Chittagong Hill Tracts. At the same time the nation has lost a significant part of its natural history. This book compiles facts, thoughts, critiques, and analyses about how the natural heritage and ecology of the CHT has been torn to pieces.

Those who have contributed to this report are SEHD research staff, academics, environmentalists, ornithologists, wildlife biologists, and students. The contributors range from highly accomplished geographer to amateur writers. But one commonality among them is they all are passionate about the ecology and the indigenous peoples of the CHT.

The book begins with a contribution from a geography guru in Bangladesh, Prof. Haroun Er Rashid, who finds the CHT as "the most romantic corner of the country" and at the same time prudently observes, "It is beset with environmental and social problems." His reflections on geology, climate, hydrography, vegetation, and environmental problems on the CHT make it clear that the CHT is very different from the Ganges-Brahmaputra Meghna (GBM) floodplains and delta, which forms the greater part of the country. He stresses on the fragility of the region and warns, "If resources of the CHT continue to be used unwisely, as at present, the environment will degrade severely in the next ten years. To prevent that bleak future every resource—trees, rocks, water, soil etc.—will have to be used in a sustainable manner," advises Prof. Haroun Er Rashid.

Other writers reinforce Prof. Rashid's contention. For many good reasons a large section of the book deals with forest, its degradation, consequences of monoculture plantations, *jum* and *jumia*. With virgin forests and forest peoples, the CHT was a majestic land even a few decades ago. Today the 'majestic natural beauty' described by Hutchinson as "very picturesque, the mixture of hill and valley densely covered with forest and luxurious vegetation" is history. Initially the British colonists and then the Pakistan and the Bangladesh governments have brought the region to a state, no less than an ecological disaster.

Writing on life in the reserved forests, Raja Devasish Roy, the chief of the Chakma Circle and a passionate researcher on the CHT, gives a scholarly account of the sufferings of the people who are made to live in appalling conditions. He and Prof. Sadeka Halim, in another chapter give an account how the indigenous peoples, despite wholesale attack on their forest and environment, still try to save the last patches of forests with traditional practices, one being the nurturing of the village common forestry (VCF).

A group of young naturalists and journalists—Sayam Uddin Chowdhuruy, Partha Shankar Saha and Buddhajyoti Chakma—have investigated the conditions of the four major reserved forests. Their reports confirm the phenomenal devastation of the reserved forest due to commercial logging, corruption, illegal felling, and militarization among others. They make no mistake in showing the immense sufferings of the people who are allowed to live in the reserved forests.

Bamboo, the stuff of life of the indigenous peoples and its phenomenal decrease during the past half century largely due to the Karnaphuli Paper Mill (KPM) has got an attention in this report. The hill people cannot think of a life without bamboo. Its drastic decrease causes miseries in their life. While Sudibbya Kanti Khisha has written on bamboo, Asfara Ahmed has looked into the environmental impacts of the KPM that is the single major factor for the drastic decrease in bamboo resources in the CHT. The gigantic mill has consumed millions of ton of bamboo since it went into operation. Asfara Ahmed has looked into other environmental concerns attached with the KPM.

The destruction of forests and the landscapes, invasion of outsiders, and promotion of plantation economy, to name a few factors, have torn the man-nature nexus in the CHT. Dr. Monirul H. Khan and Ronald Halder, two naturalists and wildlife photographers have shown why and how many flagship animals and birds have disappeared from the CHT or remain under threat. Vocal about the protection of wildlife they want to see the conservation and management efforts in this area significantly scaled up for protection of whatever is left.

With the disappearance of the forests from most part of the CHT, unknown numbers and quantities of species have also disappeared. This is calamitous both for the species and knowledge about medicinal plants. Yet Dr. Sardar Nasiruddin, in an extract from his voluminous published work, has discussed the traditional use of medicinal plants in the Chittagong Hill Tracts. His contribution is a pointer to the value of natural forest for protection of species and knowledge of traditional treatment.

Chapters on brick burning, invasion of tobacco and timber and furniture trade discuss how these have contributed to the destruction of forests, loss of cropland, wholesale destruction of biodiversity, and displacement of indigenous communities.

The people in the CHT face manifold challenges with water. Treated as gift of "Goddess" by the indigenous communities, water sources like rivers, canals, natural springs, dug wells and hilly holes in the CHT are no more abundant and safe enough. In a chapter on water, Tahmid Huq Easher, a young professional, discusses how water scarcity, its harvest and pollution cause sufferings of the people in the hills. Solutions at the local level are also discussed in this chapter. In an investigative report on stone mining Shekhar Kanti Ray and Md. Safiullah Safi show how in remote areas it causes pollution of water and contribute to even eviction of hill people from some areas.

Two interesting chapters of the book are on houses and traditional food of the hill people. Each indigenous community has its own style and precision in building houses. Han Han and A.K.M. Muajjam Hossain, two young architects have given a wonderful description of architectural designs of houses in forest villages and have shown how skilled the hill indigenous are in building their eco-friendly houses. A chapter on traditional foods simply narrated by four students in Bandarban town brings to the fore diverse and genuine cuisines different hill people enjoy every day. The outsiders, who frequent the CHT, also bring unforgettable memories of dining with forest villagers.

A very important chapter on the impacts of militarization in the CHT by Jenneke Arens, a human rights activist from the Netherlands who frequently visits the CHT, deals with harsh realities associated with militarization in the CHT. In the CHT the soldiers are omnipresent everywhere and in almost all matters. The hill indigenous peoples have memories of genocides, physical torture, rape, burning of their houses and villages, and many other forms of human rights abuses. The Peace Accord of 1997 has brought some relief but the army still sets the rules of thumb to the inconvenience of hill indigenous peoples. The temporary military camps themselves caused huge jungle clearing for their safety. The presence of the military transmits fears among the hill indigenous peoples rendering them largely passive observers of the ecological disaster in the CHT.

An annotated bibliography of books, reports and documentary films on the CHT and the glossary, concepts, and theories relating to the CHT bring useful practical tips for further research and investigation on the CHT environment.

All that has been compiled in this book is no more than the tip of the iceberg. We went close to so many stories not told in this reports for our time and resource limits. There is so much of diversity in nature and human souls to be understood; and so much of anger and inner pains among the indigenous peoples of the CHT to be felt! Yet, we believe this book comes as a practical guide to look into and understand the ecological disaster that has happened in the CHT. At the same time it comes as an eye-opener to the state and the majority who can still play the role in saving the last bits and pieces of nature and repairing the nexus of nature and human beings.

Philip Gain, *Editor*

The Chittagong Hill Tracts after the creation of Kaptai Lake

Hill Tract Region
Reserve forest
Kaptai Lake
River

1 Kasalong RF
2 Kaptai (Sitapahar) RF
3 Reingkhyong RF
4 Matamuhuri RF
5 Sangu RF

0   40 Kilometers

N

INDIA (Tripura)

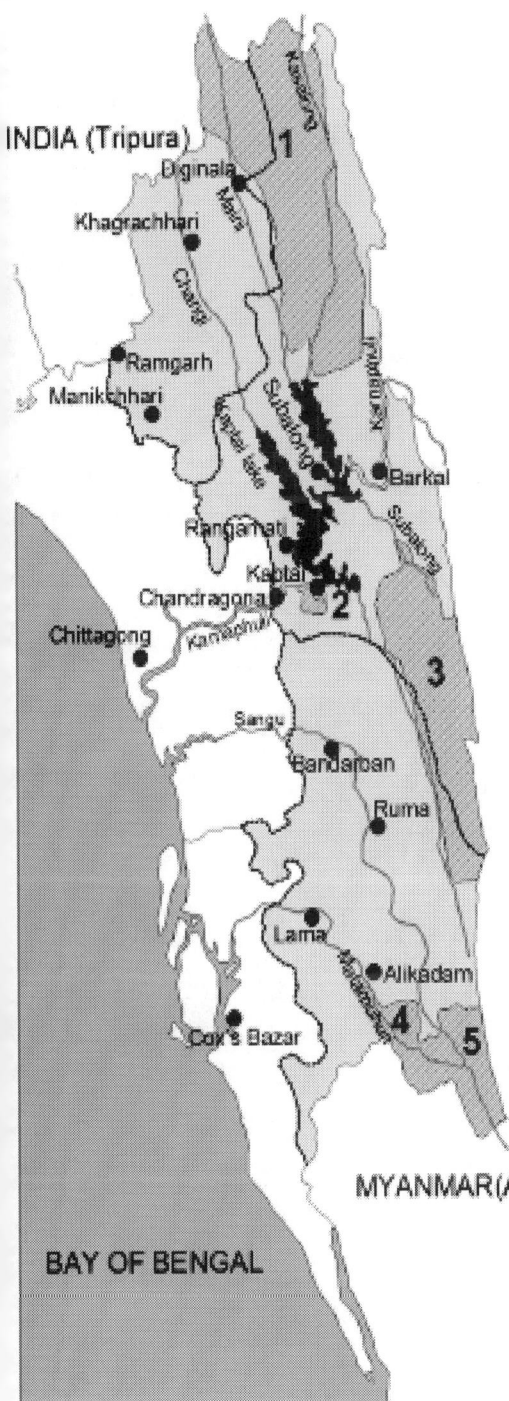

Diginala
Khagrachhari
Ramgarh
Manikchhari
Barkal
Rangamati
Kaptai
Chandragona
Chittagong
Sangu
Bandarban
Ruma
Lama
Alikadam
Cox's Bazar

MYANMAR(Arakan)

BAY OF BENGAL

Geography and Environment

Haroun Er Rashid

Introduction

The Chittagong Hill Tracts (CHT) is the most romantic corner of the country but it is beset with environmental and social problems. This side of the country is very different from the Ganges-Brahmaputra Meghna (GBM) floodplains and delta, which forms the greater part of the country. It is a country of hills and valleys with relatively little flat land. It is not quite unique because similar terrain exists in Chittagong, Cox's Bazar, Maulvi Bazar and Sylhet districts. In these areas and in the three large districts of the CHT the environmental and development problems are different from that of the plain lands of the GBM delta. These differences are largely geographical but are increased by social conditions.

The CHT is administratively composed of three districts, namely Khagrachhari, Rangamati and Bandarban. The whole of the CHT covers 13,295 square kilometers, with Rangamati containing 6116 sq. km., Bandarban 4479 sq. km., and Khagrachhari 2700 sq. km. The CHT is fully nine percent of the country. It is even more because the superficial area is not counted, which means that the area is mapped from far above and the topography is essentially treated as if it is flat. If on the other hand the actual land area of the hill and mountain slopes is measured, the surface area will be much more than the superficial area. Rough calculations show that the surface areas of Bandarban and Rangamati districts will be a fifth more and that of Khagrachhari a sixth more than the superficial area. This has an environmental aspect, because the steepness of the slopes increases the conservation problems. The other factors are the nature of the soil, the vegetation cover and even more importantly the manner of land use. The topography of the CHT creates some inherent problems for environmental conservation, and unfortunately these are exacerbated by land use practices, which are detrimental to the sustainable future of this region.

Topography

The southeast of Bangladesh is known geographically as the Chittagong Region (Rashid 1991). It is comprised of the three districts of the CHT and the similar, but coastal, districts of Chittagong and Cox's Bazar. This Region is characterized by the dominance of several longitudinal mountain ranges, which are the western ridges of the Arakan Yomas and Chin Hills. Though these high mountains reach over three thousands meters in Myanmar they are termed as hills only in comparison with the Himalayas. They should in fact be called the Chin mountains. In Bangladesh peaks and ridges above 300 meters (a thousand feet) are termed mountains or *pahar* while undulating land up to 30 meters in height are generally termed as *tilas* or hillocks, and anything higher (up to mountain height) is *mura* or hill.

The Chittagong Region, of which the CHT is a part, is a succession of mountain ranges and valleys in the east-west (latitudinal) direction. All the mountain ranges, a dozen of them with outliers, are hog-back with the

Hills in Kaptai area in Rangamati Hill District.   © Philip Gain

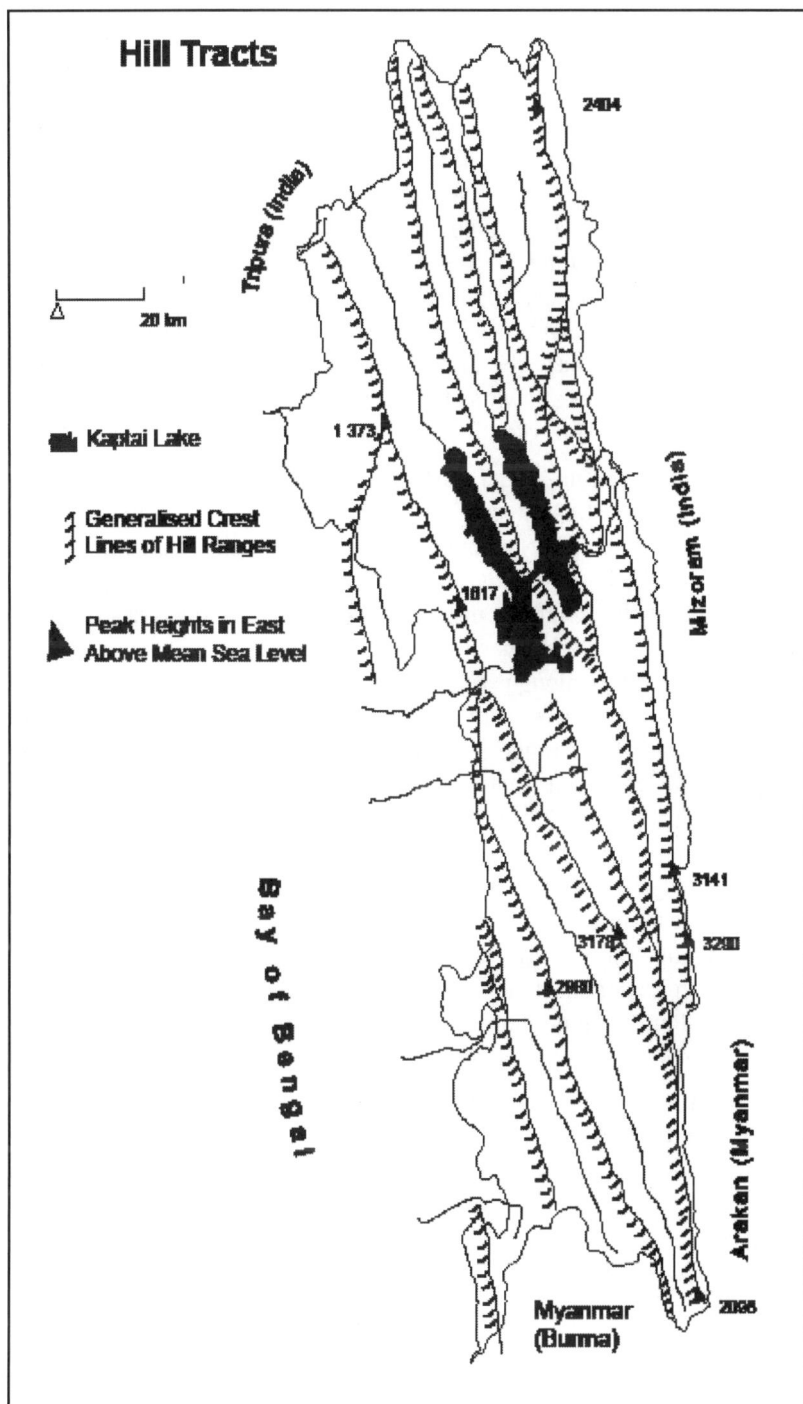

Hill Tracts

20 km

Kaptai Lake

Generalised Crest
Lines of Hill Ranges

Peak Heights in East
Above Mean Sea Level

Tripura (India)

Mizoram (India)

Bay of Bengal

Arakan (Myanmar)

Myanmar
(Burma)

2404

1 373

1617

3141

3178

3280

2980

2086

ridge lines thirty to a hundred meters wide. The Western sides fall more steeply than the Eastern sides and have scarps and waterfalls (Rashid 1991). These ridges form sharp watersheds and therefore play a very important environmental role in water retention and water distribution. The mountains north of the Karnaphuli river continue south but often bifurcate (Map). In the north the westernmost ridge is the Phoromoin range, which begins along the upper Feni river, where the range is known as the Sardengmura. Further south the town of Khagrachhari is on its eastern flank. Major peaks are Bati moin (563 m) and Phoro Moin (497 m), which tower up near the Chittagong-Rangamati road. Further south the parallel ridges are known as Rampahar and Sitapahar. They form the Shilchhari gorge and pass south of the Karnaphuli as a tangle of high hills.

The next mountain range eastwards is Dolajeri, which flanks the eastern side of the Chengi river valley in which is situated the town of Khagrachhari. This valley is about 80 km long but the average breadth is only 6 km. The highest peak on this range is Langtrai (429 m). On the eastern side of the Dolajeri range is the valley of the Maini river, which flows into the Kassalong river (often called a *khal*). Between the valleys of the Maini and Kassalong is the Bhuachhari Range, rising to 350 meters at Gilamoin and 611 meters at Changpai peak. In the southern part of the CHT the eastern-most range is the Barkal range, known as Chipui or as Lungsir by the Lushais. Its highest peaks from north to south are Khantlong ((683 m), Thangnang (735 m), Lungtian (679 m), Chipui (480 m), Bara Taung (447 m) and Barkal (572 m). This high mountain range has some branching ridges. The eastern most is the Saichal Mowdok range, which is called Saichal in the northern part and Mowdok Mual in the southern part. This range is the southern extension of the Barkal range. In the north it forks with a short arm going northwestwards. This is the Belaichhari range with Phukmani Moin (590 m) and Belaichhari peak (669 m) as its highest points.

At the junction of Belaichhari and Saichal ranges is Saichal peak (648 m). Further south the Saichal Mowdok range forms the boundary between Myanmar and Bangladesh. The peaks in this stretch are Waibung (808 m) Naphrai (748 m), Reng Tlang (958 m), Mowdok Tlang (905 m) and Mowdok Mual (1004 m).

Between the Belaichhari and Saichal ranges is the Subalong watershed through which the Karnaphuli flows into the Subalong Gorge. It is also the origin of the Subalong river, which flows west in a narrow valley to join the Reinkhyong river and into the Karnaphuli (now a lake). Further west is the headwater and deeply incised valley of the Reinkhyong river. To its west is the Politai range with Ramui Tlang (921 m), Politai (831 m) and Keokradaung (884 m) peaks. The Politai range is the southern continuation of the Phoromain. Westwards lies the narrow valley of the Sangu river flanked by a broad belt of hills and *tilas* on both sides, and the Chimbuk range to the west. This range, also known as Tyambang, continues south as Moin Daung and passes into Myanmar. The main peaks of the Chimbuk range are Lulaing (702 m), Thainkhiang (894 m), Kro (868 m), Rungrang (849 m) and Tindu (898 m). West of this range there is again a wide belt of small hills and *tilas*, with the narrow valley of the Matamuhari river winding through them. Further west is the Muranja range, which rises in the Chunoti-Harbang hills and further south merges in the hills of Naikhongchhari. Its main peaks are Muranja or Mirincha (502 m), Nashpo Taung (586 m) and Basitaung (670). At the southwest corner of the CHT is the Wayla Taung range with a peak of the same name at 414 m. Most of this range is in the Rakhine state of Myanmar.

Geology

The creation of CHT in its present physical form began in the late Cretaceous age, between 90 and 75 million years ago (mya). The mountains of CHT are a continuation of the Chin Hills, which is the northern continuation of the Arakan Yoma. This large folded area, which links the easternmost Himalayas with the Sunda Arc in Indonesia, began to form much earlier, in the Triassic Age (200-240 million years ago), but the area of the CHT was then still under the newly formed Indian Ocean. As the Arakan Yoma formed fully the folds (a succession of anticlines and synclines) extended westwards into the nascent Bay of Bengal. The shallow sea of this area belonged to the Indo-Pacific Zoo-geographical province and contained Cretaceous marine fauna similar to those found as far away as Madagascar and Natal. In the upper Cretaceous the Arakan Yoma emerged high enough as a complete ridge to separate the Burma and Assam Gulfs. The Chittagong Region began to form along the outer edges of the Yomas from Bengal Basin sediments in this period. In the words of the eminent geologist, Dr. F.H. Khan (1991) "the hilly regions of Chittagong, Hill Tracts and Sylhet district are formed by folded Basin sediments". He describes that in this region a large number of anticlines and synclines, both symmetrical and asymmetrical, are dominant. Khan wrote that most of the structures trend

NNW-SSE and that this trend may be designated as the Chittagong Strike. Subsequently in the Late Middle Miocene period (roughly 14 to 17 million years ago) severe faulting reshaped the mountain range profiles. In the Chittagong Region sediments of the Surma group forms the crest of all the anticlines barring a few, which expose the Tipam group (Khan 1991). The Surma group belongs to the Early Miocene Epoch (approx 20-25 mya) and consists of siltstone, shale and sandstone. The Tipam group is of Middle Miocene Epoch (approx 15 -20 mya) and consists of massive sandstone and some shale. According to Khan (1991) "the Tipam sandstone must be considered as a potential reservoir of gas". The valleys in the CHT are in synclines, the original rocks covered by erosional deposits and flanked by hills and hillocks of Pleistocene origin. The Chittagong region is not rich in minerals but the possibility of finding hydrocarbons is promising. Explorations should be carried out urgently since the demand for natural gas is increasing steeply. When it is exploited the supply within CHT should get priority.

Climate

The whole of the CHT falls within the Wet Tropical Climatic zonation of Koeppen (1931), because of the high rainfall even though there are five months (November to March) which are nearly rainless. However dewfall is high from October to February, and important in keeping the natural vegetation of leeward slopes evergreen. Though the western slopes of the mountains receive more rain than the eastern slopes, because the rain clouds come from the west, the long hours of sunshine on the western slope makes them ecologically drier. The average annual rainfall in Rangamati is 2500 mm and 2000 mm in Barkal, but most of the CHT receives an annual average of 2500 mm to 3000 mm. The higher ranges may be far wetter with 3500 mm or more. The Rangamati area east to Barkal receives the least rainfall, but it is sufficient to keep the soil moist wherever there is sufficient tree cover.

The temperature regime is very mild, allowing wet tropical and even equatorial plants to be grown in the valleys. In Rangamati the mean monthly maximum of 35.1 degree Celsius is reached in April, but in January the mean maximum is 26.4 degree Celsius. This is the coolest month with mean minimum at 10.2 degree Celsius. Rangamati is at a low location and the proximity of the Lake reduces the cold. On the mountain ranges the effect of height and more exposure to northerly winds brings temperatures down by five to ten degrees during cold waves.

It is important to note that the western side of the mountain ranges is directly exposed to the cyclones, which come up the Bay of Bengal. Wind speeds can reach over 100 km per hour because the seacoast is merely thirty miles west of the Muranja and Tyambang mountains.

The climatic factors (rainfall, dewfall, diurnal temperatures, sunshine hours, wind speeds and wind directions) have not been adequately recorded in the different valleys and mountain ranges. This work needs to be undertaken because climate is a very important component of the environment. With ongoing climate change it is urgent that more data on present climate and its trend is made known and used for development purposes.

Hydrography

Due to the abundant rainfall and the soft soils the topography is etched with innumerable streamlets (*jiri*) and streams (*chhari*). The drainage pattern is essentially that of a trellis. Twenty to thirty first-order streamlets augment the flow of every second-order stream over a hundred of which flow into every river. For many purposes the effect of environmental factors can be summarized by the drainage density of a river's catchment area. Hardly any work has been done to ascertain the drainage density of any part of the CHT. It may be seen that the third-order drainage pattern is quite regular and superimposed on a topography, which developed after the main rivers had already been established. It is evident that the Karnaphuli (Kynsa Khyong) is an antecedent stream, flowing through a terrain that can be described as a scarp-and-vale landscape. This river that rises in south Mizoram (India) makes a typical elbow turn before entering Bangladesh and cutting through the Subalong Range by the Silchhari gorge. In India the Tuilianpui is a major tributary, while within Bangladesh the Thega Reinkhyong, Subalong, Kaptai and Ichamati are the main tributaries on the left bank, and Sajek, Horina, Kassalong, Chengi and Halda are the main tributaries on the right bank. The estimated annual discharge of the Karnaphuli river is 25 million acre-feet. These tributaries flow between the ranges and are not antecedent. The Kaptai dam was constructed on the Karnafuli river in the 1950s to generate electricity. Its capacity of 60,000 kw was a big amount at that time but has since dwindled to only about one percent of the nation's requirement. It forms large lake (reservoir), which is 268 sq. km. in size in the dry season and 742 sq. km. in the rainy season. Not only does the wet and dry season affect its extent but also the needs of power generation. Large discharges may be made in the wet season to keep the lake surface at a safe level. This can create erosion and flooding downstream. On the other hand the level may be kept high after the rains thus delaying rice transplanting in flooded areas of Chengi and Kassalong valleys. The prime consideration is power generation and not food production. Now that the amount of power generated has become a very tiny fraction of the national output thought should be given to local needs in the CHT. This lake drowned the middle Karnaful valley from Kaptai to Barkal and the lower reaches of seven tributaries. At the time of its formation half of the lowland paddy of the CHT was grown in this area and all of it was lost. Subsequently the area exposed by the receding lake in the dry season has been developed for plough cultivation. Nevertheless the Kaptai Lake has had a profound effect on the geography, society, economics, politics and environment of the CHT. The two other major rivers in the CHT are the Sangu and Matamuhari (Mamori). The Sangu, also known as Shonkho or Rigre Khyong, rises in the Sangu National Park, between the Politai and Chimbuk mountain ranges. It has small tributaries within the CHT, namely Dolu, Hangor and Tankabati. The estimated annual discharge of the Sangu is about 5 million acre-feet. Further south is the Matamuhari or Moree Khyong, which rises in hill country between the Muranja and Chimbuk ranges. It has small tributaries within the CHT. The annual discharge of the Matamuhari is only three million acre-feet. Two other rivers deserve mention. The Feni rises in Sardeng and forms the boundary between Bangladesh and the Indian state of Tripura for much of its course. The left bank is entirely in Bangladesh and receives the flow of many *chharas* draining the northern Phoromoin and the hills of Manikchhari, such as the Pilak chhara. In the southern extremity the Bagkhali river rises in Naikhongchhari thana and flows into the sea at Cox's Bazar.

The hydrogeology of many places in the CHT, in particular in Bandarban Hill District, is being severely affected by stone quarrying. This is not being done by local people but by contractors from outside with the connivance of the local political apparatus. By removing the stone layers the streams (*jiri* & *chhara*) are drying up. This disastrous consequence has led to the relocation of many *paras*. If this continues drinking water supply will dwindle, run-off will increase and the soils of many areas will become doughtier. Even at present food production is barely adequate, so that further deterioration may lead to localized famines or the creation of environmental refugees. It is recommended that indiscriminate stone quarrying should not be allowed and that water supply for existing *paras* should receive the highest priority.

Vegetation

The natural vegetation of the CHT used to be lush in areas not touched by *jum* (swidden) cultivation. On slopes, which were jumed and then put under long fallow the growth was of course secondary, with many of the shade-loving plants unable to establish themselves. At present the natural vegetation areas have greatly shrunk and the other areas are affected by the introduction of exotics or the spread of invasives. There are two reasons for this situation. One of the reasons is the increase in demand for *jum* land due to the natural growth in population, and the other reason is the planting of exotics by the Forest Department to meet pulpwood demand by the Karnafuli Paper mills.

On both sides of the upper Maini and all along the Kassalong khal, between the Bhuachhari Moin and the Chipui Tlang, is the Kassalong Reserve Forest, the second largest in the country after the Sundarban. On paper it is 3,384 sq.km. in area but in reality much less because of legal and illegal settlements. In the west central area is the Kaptai R.F. straddling the Rampahar-Sitapahar hills. South of the Karnaphuli river and on the eastern side are the Subalong and Reinkhyong R.F. Further south are the Sangu R.F. and the Matamuhari R.F., created to protect the headwaters (watersheds) of the two rivers. Whereas the Matamuhari R.F. has been partly denuded the Sangu R.F. is in pristine condition and has been declared a National Park.

The forests are of two types, tropical wet evergreen and tropical mixed deciduous and evergreen. The Kassalong R.F. is dominated by Garjan (Dipterocarpaceae) species and is therefore largely mixed evergreen. On the eastern sides of the slopes, particularly in the upper reaches, one can find evergreen forests with Chapalish (Artocarpus chaplasha), Chundul (Tetrameles nudiflora) and Telsur (Hopea odorata) in the first storey, Pitraj (Aphanamyxis polystachia) and Toon (Cedrela toona) in the second storey and Horina (Panicovia rubiginosa) etc. in the third storey. This is essentially the mix of forest types in the various Reserve Forests. In many areas there is a thin covering of trees, consisting of Civit (Swintonia floribunda), Chikrasi (Chukrassia tabularis), Banderholla (Duabanga sonneratoides) etc. These areas are often protected by the local people as their "mouza forest". Efforts should be made to increase *mouza* forests and give them strict legal protection.

A significant part of the hillsides consists of large bamboo brakes. Seven varieties occur, of which Muli (Melocanna baccifera) and Mitenga (Bambusa tulda) are the commonest. The flowering and subsequent die-back of Muli bamboos once every thirty years denudes the hillsides of these brakes, spurs an increase in rodents and seriously affects the food supply of the *jumias*. Bamboo is used as raw material by the Karnaphuli Paper Mill (KPM) in Chandraghona, but its shortage for various reasons has encouraged the use of fast-growing trees, which in turn has led to plantations in areas formerly *jumed* and thus increased local food shortage. Another way in which investors from other areas benefit from the CHT but give little to the locals is through rubber plantations. Lately large areas in Naikhongchhari have been planted with rubber and many locals have been deprived of land.

In the *jum* (swidden) hill-rice, maize, cotton, oilseeds, and vegetables are grown. They benefit from the leaf residues and ash left from burning the slopes. It is surmised that soil erosion is heavy but actual measurements of soil loss are very scanty. More research into this aspect is urgently required. Tobacco has spread very rapidly, particularly in the areas with richer soils,

A magnificent stream in the CHT.   © Philip Gain

such as the upper Matamuhari valley. In the absence of organic manuring the long-term effects will be adverse. As far as the social environment is concerned the increase in tobacco use will lead to greatly increased expenditures on health.

The government had introduced and encouraged fruit cultivation since the 1960s. At present the CHT produces large quantities of bananas, pineapple and jackfruit, and thousands of tons are sent to other parts of Bangladesh every year. Cashew was introduced and for some years it was a promising crop, but lack of agro- management has severely dwindled its production. Tree crops are generally favorable for the hillside, if they are not monoculture plantations. The growing of exotics for KPM or rubber for industries elsewhere are detrimental to the natural vegetation and should be curbed.

Lowland rice is confined to the valley bottoms and to the edges of Kaptai lake when its level falls. Once there were plenty of marshes but they have mostly been converted to rice lands. Nevertheless there are many small areas where one can find Kainjal (Bischofia javanica), Jarul (Lagerstroemia speciosa) and Jaitbet (Calamus viminalis). These spots add greatly to the biodiversity of insects and herbaceous plants, and therefore they enrich the environment.

A rare patch of surviving evergreen forest in the CHT.   © Philip Gain

Environmental Situation

The CHT Soil and Land use Survey by FORESTAL Ltd (Vancouver 1967) found that only 3.3% of the total land area in CHT was good for lowland crops. These lands are essentially valley bottoms, mostly along the larger rivers. It was also found that 19.0% of the land is suitable for arboriculture or dry-foot crops such as cotton, *jum* rice, vegetables, oilseeds etc. These lands are largely *tilas* and hills in-between the larger ridges.

The remaining 77.7% of the land is suitable only for tree crops, that is fruit, nut, timber, firewood species. Many years have passed since that survey and population pressures have increased manifold, often leading to unsustainable land use or even irreversible damage. A detailed land use and land capability survey is very necessary right now, because of growing land disputes and growing degradation of the natural resources.

Despite its outward appearance the environment of the CHT is fragile. The resources of this beautiful land are not being used wisely. If these resources continue to be used unwisely, as at present, the environment will degrade severely in the next ten years and adequate productivity for the people will become unsustainable. To prevent that bleak future every resource—trees, rocks, water, soil etc.—will have to be used in a sustainable manner. This is not being done at present. There has to be more consultation with local people and a change in the attitude of development agencies and the administration. A cardinal principle has to be that any change or development, which is not beneficial to local people should not be carried out.

References

Adnan, Shapan (2004): *Migration, Land Alienation and Ethnic Conflict: Causes of Poverty in the Chittagong Hill Tracts of Bangladesh.* Research and Advisory Services, Dhaka

Arannyak Foundation (2010): *Promoting Alternative Livelihood for Forest Conservation.* Dhaka

Asiatic Society of Bangladesh (2003): *Banglapedia.* Online. http:/ banglapedia. search. com.bd/

CHTDF (2009): *Socio-economic Baseline Survey of Chittagong Hill Tracts.* Prepared by Human Development Research Centre for CHT Development Research Centre for CHT Development Facility; UNDP, Dhaka

FAO/UNDP (1988): *Agroecological Regions of Bangladesh; Land Resources of Bangladesh;* Land Resources Appraisal of Bangladesh for Agricultural Development; Rome.

FORESTAL (1967): *Chittagong Hill Tracts Soil and Landuse Survey 1964-66* (9 vols). Forestry and Engineering International Ltd., Vancouver.

Gain, Philip (2006): *Stolen Forests.* SEHD, Dhaka

Gain, Philip (ed) (2000): *The Chittagong Hill Tracts: Life and Nature at Risk.* SEHD, Dhaka

GOPRB (1975): *Chittagong Hill—Tracts, Bangladesh Gazetteers.* Ministry of Cabinet Affairs, Establishment Division; Government of the People's Republic of Bangladesh.

Khan, F.H. (1991): *Geology of Bangladesh.* University Press Ltd., Dhaka

Khan, M.S. (2011): *Strategies for Arresting Land Degradation in Bangladesh* in *Strategies for Arresting Land Degradation in South Asia.* SAARC Agriculture Centre, BARC, Dhaka.

Krishnan, M.S. (2001): *Geology of India and Burma.* (CBS Publishers, Delhi)

Rashid, H.E. (1991): *Geography of Bangladesh.* University Press Ltd., Dhaka.

Rashid, H.E. (2005): *Economic Geography of Bangladesh.* UPL, Dhaka.

The Chittagong Hill Tracts: An Ecological Disaster

Philip Gain

Isolated from the outside world for centuries the Chittagong Hill Tracts (CHT), at one time, was a majestic land. The entire region was covered with dense forests of rich flora and fauna. At 13,295 square kilometers, or 10 percent of Bangladesh, the area is mountainous that greatly contrasts with the rest of Bangladesh in geography and human habitation. This Southeastern part of Bangladesh, the CHT, is surrounded by the Chittagong and Cox's Bazar plains stretching along the Bay of Bengal on the west, by the Indian states of Tripura and Mizoram on the north and east and by the Arakan region of Myanmar (Burma) on the south and southeast. The region really forms a bridge among Bangladesh, Myanmar and India. While most of the country is flat and a few feet above the sea level, the CHT is completely different in physical features, landscapes, agricultural practices and soil conditions from the rest of Bangladesh. Plough cultivation, which is a common feature in the plains is seen only in the plains patches in the mountain valleys. The terrace farming, a common agricultural practice in the mountains is also not seen in the CHT hills, the highest of which is close to 4,000 feet.

Its 'majestic natural beauty' had been described by Hutchinson as 'The scenery throughout the District is very picturesque, the mixture of hill and valley densely covered with forest and luxurious vegetation, yields the most beautiful and varied effects of light and shade. To be viewed at the best it should be seen from the summits of the main ranges, where the apparently boundless sea of forest is grand in the extreme. The cultivated area of the valleys, dotted here and there, appear as islands, carpeted with emerald green, cloths of gold, or sober brown according to the season of the year. The rivers slowly meandering on their way to the sea, now

Seen from Mong U Para, a Khumi village in Roangchhari Thana in Bandarban Hill District, a rare forest patch on a winter morning. © Philip Gain

shimmering like liquid gold, and again reflecting in heavy dark shadows every object within reach, all combine to make a picture not easily forgotten' (Hutchinson, 1909, 2 in van Schendel et al., 2000:121).

The description that Hutchinson had given a century ago about the CHT still makes sense to a certain extent although human interference has caused enormous damage to this beautiful region. One can be charmed by the spectacular scenic beauty of the Kaptai Lake as one approaches the Rangamati City, but this "beauty could not hide the fact that it was catastrophe for thousands of Hill cultivators" (van Schendel et al., 2000:143). Then one will be shocked to see the army camps on the shaved mountaintops and troops patrolling the streets even after the peace accord has been signed in 1997. The presence of the military in every corner of the CHT not only indicates the massive human rights abuses throughout the past two and half decades [during insurgency time], but also the ecological disaster, which is evident in the hill slopes and the ground where little of the original forests survive today.

The CHT also was and still is to a large extent a land of unique indigenous peoples, natural history and resources, education, and ecology that sustained unique life forms and cultures. Today with the strong presence of the military all around and the Bangalis settlers protected by the military have made the life of the indigenous peoples miserable and natural resources severely exhausted. An ill-conceived argument to justify the Bangali settlement is that around 135 persons live per square kilometer (according to 2001 census) in the CHT compared to the national average of more than 1100 persons. This is indeed a myth that there is vast unused land in the CHT. In fact, only 2% of the CHT land is suitable for wet-rice cultivation. The bulk of the land (51%) is good only for forestry, while the rest of the CHT can support horticulture, *jum* (shifting) cultivation, and some terrace farming.

Land use situation of the Chittagong Hill Tracts

Land Use	Acres	Percentage
Land suitable for rice cultivation	77,000	2
Land suitable for horticulture and crops	670,000	21
Land suitable for forest only	1,600,000	51
Reserved forests	800,000	26
Total	**13,147,000**	**100**

Source: Based on Forestal's survey, 1964 to 1966 in Anti-Slavery Society (now Anti-Slavery International) London, 1984. *Chittagong Hill Tracts: Militarization, oppression and the hill tribes.*

The sloping land used for *jum* cultivation traditionally belonged to different indigenous communities. Exclusive individual rights to *jum* plots had never been established, and community members could claim ownership only over the crops grown on the plots. The British colonizers exploited the communal land use arrangement in the CHT to establish "supreme and unlimited authority" over the land. Thus did the CHT land come under direct state control. Henceforth, the hill slopes became the property of the state. The hill people continued to engage in shifting cultivation but they were levied a *jum* tax to encourage them to shift to sedentary agriculture.

Land, Development, and the Environment

The state-sponsored and aid-dependent 'development' initiatives that the Pakistani Government had started after the independence in 1947 actually began the post-colonial process of the devastation of the CHT land and massive environmental degradation. During the Pakistan time, two "pleasant" development interventions of the state were Karnaphuli Paper Mill (KPM) and the Kaptai Hydroelectric Project that equated to "national pride". But the projects turned out to be causing unprecedented ecological damage right away and posed the biggest ever threat to the hill indigenous peoples and their economic, aesthetic, and political life.

The Karnaphuli Paper Mill (KPM): Financed by external resources (US$13 million) including a World Bank loan of US$ 4.2 million (Arens 1997:49 in Bhaumik et al., eds) the first large-scale development project of the CHT, the huge mill, KPM, is situated at Chandraghona, 26 miles upstream of Chittagong. It is also a convenient location to connect the mill to the Chittagong Port. It went into operation in 1953. To begin the work of the giant industrial unit, "the hill began to be leveled, deep ravines and crevices filled, thick and dense forest cleared" (*Pakistan Moves* 1956, 2 in Schendel, Mey and Dewan leveled1).

The mill has been polluting the waters of the Karnaphuli River since its establishment. One crossing the Karnaphuli River at Chandraghona is almost certain to view a few mile long column of the foam floating in the Karnaphuli River. This is created from the wastewater dumped from

The huge Karnaphuli Paper Mill at a turn of the Karnaphuli River at Chandraghona.   © Philip Gain

the Karnaphuli Paper Mill directly into the river. If one takes a look at the water, one will also find the waste particles mixed with the water making it colored. Local people complain they sometimes see dead fish floating near the paper mill. This can be the effect of toxic dioxins, which are normally found in the downriver of the pulp mills. There are also scientific evidence that pollution from the paper mills can change the sex organs and other features of fish swimming within waters close to paper mills. Research by the National Water Research Institute in Burlington, Ontario and the Pulp and Paper Research Institute of Canada indicated in 1992: "bleach-free mills were as dangerous to fish as those emitting organochlorines from bleaches" (Marchak 1995:46). The people living close to the Karnaphuli Paper Mill also complain of foul odor that the mill sends into the air. Any attempt to reduce the pollution of the Karnaphuli River was likely to fail because of the poor environmental standards in Pakistan and then in Bangladesh.

The raw materials for the Karnaphuli Paper Mill are mainly bamboo and softwood. The mill was given the right for 99 years to extract these raw materials from the forest. It extracted millions of tons of bamboo and softwood from the hills to keep the mill running. However, the KPM that became an important icon of economic development for Pakistan set forth the conditions for an environmental catastrophe in the CHT and misery for the hill people. The bamboo and softwood from the native forests is now so decreased in the CHT that there is perhaps no alternative to industrial plantation, which severely limits the customary land rights of the hill indigenous peoples.

What is concerning now is that more pulp and paper mills are being established, which will look for raw materials from the hills and industrial plantations. With the natural stocks drastically reduced the local communities now see with great astonishment the plantations (which are agricultural crops, not forests) with exotic species that are taking the place of the natural forests. Plantations for supplying raw materials to primarily KPM, once a top industrial showcase of Pakistan, have caused extensive deforestation, destruction of species and ruined much of the *jum* lands to the detriment of the hill indigenous peoples.

The KPM that became an important icon of economic development for Pakistan thus set forth the conditions for an environmental catastrophe in the CHT and misery for the hill people.

The construction of the KPM created 10,000 jobs, but the hill people got only around five percent of these and mainly low-ranked jobs (Arens 1997:49 in Bhaumik et al., eds). The same story was repeated in the construction of the Karnaphuli Rayon Mill in 1966.

The Kaptai Hydroelectric Project: A few years after the construction of the Karnaphuli Paper Mill, the Pakistani symbol of development, the Kaptai Hydroelectric Project, was put into operation. Starting in 1959, the US$100-million project was completed in 1963 (van Schendel et al., 2000:203). The hydroelectricity produced at Kaptai thereafter supplied the energy needs of the capital Dhaka and of Narayanganj City. The electricity also became crucial to run the wheels of industry.

What became a great triumph for the Pakistani rulers had catastrophic effects on the hill indigenous peoples and the environment. It created a 650 square kilometer upstream reservoir, submerged 40% of the most productive valley land of the CHT, many villages, and forests. It displaced around 100,000 people who constituted one-quarter of the region's population [Gain (ed.) 2000]. Many of those displaced had no other choice left but to push up into the hills and take to *jum* cultivation, which they had abandoned a long time ago [Gain (ed.) 2000]. The project also hugely

Aerial view of Kaptai town and dam. Courtesy: Sandercock, 1964-65 in Schendel, Mey & Dewan, 2001, The Unversity Press Limited.

contributed to people's internal displacement. Unsurprisingly, the project generated discontent and anger among the indigenous peoples in the CHT.

Aside from the immediate damages such as inundating croplands, villages, and forests, the Kaptai hydroelectricity dam and the large lake that it created had far reaching ecological effects. Because the reservoir inundated many of the best forested valleys, most of the wildlife that once comprised of bison, sambur, barking deer, leopard, Royal Bengal Tiger, panther, etc. is not seen anymore. In less than forty years after the construction of the Kaptai dam the tiger species have gone totally extinct in the CHT. The elephant population has drastically decreased. Although the CHT still charms the tourists and outsiders, many now term it as a hilly park devoid of any significant wildlife. The water pollution in the Kaptai Lake has become alarmingly hazardous. Most of the people living around the Kaptai Lake depend on the lake water for bathing, washing, cooking, and drinking. Human excreta and the residue of pesticides and fertilizers, which make their way into the lake every day pose health risks. Water pollution has also severely affected the life in the water.

No social and environmental impact assessment of the Kaptai

Kaptai Lake that has submerged 40% of the most productive valley land in the entire CHT.   © Philip Gain

Hydroelectricity Project was done prior to the implementation that would inundate huge areas of the Hill Tract and displace a quarter of the entire population of the region. However, it was only three years after the implementation of the project and the colossal damages being done to the ecology and the indigenous peoples that the Government of Pakistan engaged a Canadian company, Forestal Incorporated to study development prospects in the CHT and the impacts of the dam on the ethnic communities. The study team comprised of 11 geologists, economists, agronomists and biologists (Anti-Slavery Society 1984: 32). It is important to note that according to L.G. Löffler the Forestal people were natural scientists who did not bother much about social or cultural questions and who thought in terms of producers, not in categories of hill people and plains people.

The study found that "the Tribal people had attained a reasonably satisfactory way of life adequately adjusted to the limitations imposed by the physical environment" (Anti-Slavery Society 1984: 33). The hill indigenous peoples were *jum* cultivators. After the construction of the dam and the creation of Kaptai Lake, they lost a large living area and explored fresh areas for their traditional agriculture. They were not responsible for the shrinking of land available for shifting cultivation. But the Forestal Incorporated blamed their traditional practice of agriculture. "The practice of shifting cultivation or jhuming ... has been widespread throughout the area. This has resulted in destruction of the original forest cover and most of the area has not reverted to forest but is now covered with sungrass and scrub. The necessity of preserving the forest was recognized many years ago and forest reserves were established over approximately 26 percent of the total area of the District." (Forestal, Volume 5, 1966: 1).

The Forestal team strongly recommended that *juming* be controlled and horticulture be extensively introduced "which they thought would bring hitherto unknown prosperity to the inhabitants in danger of pauperization" (Löffler 1991: 7).

It was insane to blame the hill indigenous peoples and *jum* cultivation and award credit to the creation of forest reserves. Because it was not for *jum* cultivation but for ill conceived development plans and other factors that led to further destruction of the CHT forests and its ecology. The newly introduced horticulture also proved to be a flop to a large extent.

After independence, no significant development plan was initiated until the time of General Zia who declared in 1976 that the problems of the CHT originated from underdevelopment. He therefore founded the Chittagong

Hill Tracts Development Board (CHTDB) in 1976 by an ordinance to solve the CHT problems through large-scale development programs. The major development interventions in the CHT since then were designed, managed or overseen by the CHTDB. It implemented projects and programs for construction of roads, telecommunication, electrification, and moving the hill people into the 'model' or 'cluster' villages, which has severe negative impacts on proper land use in the CHT.

The roads have been beneficial for transporting produce to markets, but they were first used to facilitate troop movement. Roads helped the military to combat the *Shanti Bahini* as well as the businessmen most of whom were Bangalis. The cluster villages were intended to isolate the *Shanti Bahini* and cut their supplies. Even after the peace accord of 1997, there are still several hundred non-permanent military camps spread throughout the CHT. These camps themselves have occupied much of the CHT land previously maintained by the hill indigenous peoples. The military camps are a major threat to land and environment, among other things.

The government policy and actions regarding Bangali settlement in the CHT have steadily reduced the resources that the indigenous peoples have accessed for centuries. These contribute to competition for land, limit the access of the indigenous peoples to forest and land, and cause severe human rights abuses. During partition of India in 1947 the Bangalis comprised 2.5% of the CHT population, which rose to 48.57% in 1991 (1991 population census) from 10% in 1951 and 35% in 1981 (Schendel 1992:95). The Bangali population jumped further up with an ill-planned in-migration of Bangalis into the hills. In one decade from 1981-1991 the CHT population increased by 67.95% or 6.79% per annum compared to 2.17% of the national increase (Gain 2000).

It is obvious that around 400,000 Bangalis had been settled in the early eighties for political reasons to bring the Bangali population in the CHT close to that of the indigenous population and to alienate the hill people from their land. The target of the state-sponsored Bangali settlement has been fulfilled to a great extent.

An amendment to the Rule 34 of the CHT Manual in 1979 lifted the restriction against the settlements of the CHT lands by non-residents. Following the amendment the Bangali settlement had begun under a clandestine plan. In the first phase 25,000 Bangali families were settled by the end of 1980. Each of these families was to be given five acres of land and a cash amount of Taka 3,600 and provisions to support them for the first few months (Mohsin in Gain (ed) 2000). Each of the families to be settled in the second phase was to be given either 2.5 acres of plain land or four acres of plain and bumpy mixed land, or five acres of hilly land, since most of the prime land had already been distributed among the first batch of settlers (Mohsin in Gain (ed) 2000). In the third phase in 1982, 250,000 Bangalis were settled. The settlement of around 400,000 Bangalis was completed by 1984.

The Bangalis had been settled on the argument that vast tracts of land in the CHT were lying empty and the government would settle the Bangalis on *khas* land. But in reality cultivable lands in the CHT were very limited. It was even before the construction of the Kaptai Dam that lands available for cultivation were insufficient for the CHT population, which were 385,079 in 1961 (Mohsin in Gain (ed) 2000). Then the reservoir created by the Kaptai Dam, which submerged 40% of the CHT's prime cultivable land, aggravated the situation.

By the time the forced in-migration of 400,000 Bangalis occurred there was actually no land available to keep the promise that the government made to the settlers. So the alternative left was to give up reserve forest and uproot the indigenous persons from their land. That is exactly what the government did. It reserved a few thousand acres, which was worth one-tenth of the requirement. Then serious abuses against the hill indigenous peoples occurred. Around 100,000 hill indigenous people were ejected under the settlement policy from their traditional land. They became homeless. About half of them became refugees in India and half got scattered throughout the CHT to become internally displaced persons (IDP). The Task Force for Refugees in the CHT made its final list of IDPs public according to which there were 1,283,643 IDP families in the CHT as of May 2000 of which 90,208 were 'tribal' and 38,156 'non-tribal'.

The government policy of Bangali settlement has had far reaching consequences for the land and indigenous peoples of the CHT. The land in the CHT simply cannot sustain the population that it now has. Both ethnocide and ecocide are evident because of the demographic imbalance and the land use policies of the government.

Despite this gloomy picture with the land the story of indigenous peoples losing land and the destruction of the web of nature and humans due to greed from the humans of the plains seem unending to date.

Land Grabbing and Destruction of Environment for Rubber and Tobacco in Bandarban

Philip Gain

Cha Kra Aung, 60, stands on the land that was in his possession till 2008. He shows exactly where he had his home. He points his fingers to the land by the side of the stream where he used to grow vegetables. Now in the month of February (2011) it is covered with luxuriant green tobacco.

Aung with his family of wife, a son, and four daughters lived in this remote forest village, Longodujhiri (Khal) Chak Para till 2008 when the last Chak families abandoned this village. Now the outsiders, mostly Bangalis, have taken control of the village land and are growing tobacco for the British American Tobacco (BAT).

Longodujhiri (Khal) Chak Para, now deserted of human habitation, is located in Baishari Union in Naikhongchhari Upazila in Bandarban Hill District. Quite unknown even to regular trekkers to the Chittagong Hill Tracts (CHT), Baishari is one of four unions in Naikhongchhari Upazila with Chak inhabitation. The Chak is one of the very small indigenous communities in the CHT with a population of around 3,000—all concentrated in Naikhhongchhari Upazila. There are another four to five thousand of them in Myanmar. Longodujhiri (Khal) Chak Para was one of seven Chak villages in Baishari Union. Now the village has vanished from the list.

"We, 20 Chak families, lived in the village for decades. Our forefathers, *jumias* (slash and burn or swidden cultivators), developed this village. The Bangalis have grabbed the land that we once had houses on and cultivated to grow vegetables," says Aung. "We all were engaged in *jum* cultivation in the hills around us. On the precious flat land (narrow straps) by the streams we grew potato, ginger, chili, mustard, radish, and some other crops. Collectively we, 20 Chak families, had some five acres of cropland in our possession."

A Chak woman who once grew staples and vegetables on this precious valley land in the hills, has now become a laborer for tobacco cultivation.

Hilly and bumpy land, commons for the use of the indigenous peoples, was not specified for swidden agriculture. But the Chaks had plenty of it around their village. In addition to swidden agricultural land in the hills and flat land along the streams, the Chak villagers had designated groves named, *Badurjhiri Charra*, to harvest bamboo and wood from. Such designated groves, these days titled, village common forest (VCF) by many, have existed in the CHT for a long time. The forest villagers collect their household necessities of bamboo and wood from these groves. They also earn extra cash from the sale of the surplus.

Thus life was simple for the Chaks of Longodojhiri (Khal) Chak Para completely dependent on land and forests for which they never had titles. "We were a well-fed and happy people with our life completely dependent on land and forests. What we grew on cropland and in *jum* was more than enough for round the year. We never thought about papers for our land and forests," says Aung.

The trouble began when the desperate outsiders, began to invade the area for bamboo and timber on land that belonged to nobody on paper. It was to the advantage of the outsiders that the constitution and laws in Bangladesh do not recognize the customary rights of the indigenous peoples in the Chittagong Hill Tracts (CHT) and elsewhere in Bangladesh.

The Chaks of Longodujhiri also had terrible times with the insurgency war fought between the indigenous peoples and the Bangladesh military for two and half decades. The Chaks participated in the insurgency war that the indigenous peoples fought for autonomy and self-rule. In 1981-1982 the Bangladesh military evacuated all Chak families from Longodujhiri (Khal) Chak Para and settled them in Baishari. They were alleged to be feeding and sheltering the members of the *Shanti Bahini*, the armed wing of Parbattya Chattagram Jana Samhati Samiti (PCJSS), a regional party of the indigenous peoples of the CHT that led the insurgency war. The families moved back to this village in 1987-1988 only to run into further troubles.

Aung says he had in his possession about an acre of land for homestead and to grow vegetables and other subsidiary crops in addition to harvest from the *jum*. All his family members were engaged in *jum* cultivation. "I am a *jumia* from my childhood. My daughters were also *jumia* with me. All daughters are married now. Three of them, married in the forest village *Hatirdatjhiri*, are still *jumias*."

A Chak man of Baishari shows the commons in the past, now under rubber cultivation. © Philip Gain

The Bangalis from different places of Bandarban Hill District and those settled permanently (also Bangalis) scaled up the harvest of bamboo and wood from *Badurjhiri Chhara* in the recent years. What was once reserved only for the Chaks of Longodujhiri (Khal) Chak Para who harvested bamboo and wood in a calculative manner, got quickly exhausted.

The biggest hassle for the Chaks of Longodujhiri came from the loss of land available for *jum* agriculture, the main source of their living. It so happened due to rubber plantation, which has quickly expanded into the hills where the Chaks and a few other indigenous communities (Marma and Mru) live.

The vegetation on the steep hills in the West and the Northwest of the Longodujhiri (Khal) Chak Para has been fresh cut. Aung says that rubber saplings would soon be planted there. Most of the land available to the Chaks for *jum* had already been invaded by rubber cultivation.

Losing all means of livelihood, the Chaks of Longodujhiri began to evacuate the village from 2003. Aung and a few others were the last families to desert the village in 2008. Aung now lives in Baishari Upar Chak Para and works as a day laborer at the rubber plantation. He still goes to the woodland that has been disappearing fast. "My economic condition is very bad. Now I have my house on just two/three decimals of land."

Thowai Cha Uk (55), Chaila Mong (40), and Mong Shaithowai Chak (32) who once lived in the village stood on their homestead only to grieve. They told the same story of deception.

Thowai Cha Uk (55) lived in Longodujhiri with his wife, three sons, and three daughters. He evacuated from Longodujhiri in 2008, with Cha Kra Aung. They settled in Longodujhiri Chak Para after years of *jum* agriculture.

"I had 40 decimals of land including the homestead in Longodujhiri. I settled in this village when I was around 15 years of age," says Uk, who after leaving Longodujhiri has settled in Upar Chak Para on a tiny piece of land. "Now I am a night guard at a rubber garden. I do not like this work. In the past I used to give work to others. Now I work for others as a laborer." Two daughters of Uk are married and remain to be *jumias*—one in Naikhongchhari Mouza and the other in Hatirdatjhiri Para in Koenjhiri Mouza. The sons work in a garments factory in Dhaka.

Chaila Mong left Longodujhiri in 2004 and settled in Baishari. "Now I am a day laborer in a rubber garden," laments Chaila.

Mong Shaithowai Chak left Longodujhiri in 2007. Now he has a place in Baishari Headman Para. He stood on his homestead where a tobacco cultivator from outside was building a house. He pointed his fingers to the land where he once grew vegetables and other crops. Helpless, he was saying in anger, "We can no more carry out *jum* and harvest bamboo. The headman has given the land to the outsiders for money."

Aung and others do not know the people who have actually grabbed their homestead, land for vegetable gardening, and orchards. All that they know is their land has gone out of their control and to the Bangalis from outside.

Mohammad Yusuf (16), a worker at tobacco plantation at Longodujhiri told on 3 February 2011, "My uncle Momtaj from Garjania Bazar (in Ramu Upazila of Cox's Bazar District neighboring to Bandarban Hill District) has taken this land (that belonged to Cha Kra Aung) from the headman of Alekhyong for Taka 10,000 (US$140)." For Cha Kra Aung (60) this is an exposé that makes him angry. "We want to get back our land," demands Aung.

The last Chak village in the east of Baishari Union is Badurjhiri. It is a half-an-hour walk from Longodujhiri Chak Para along a stream that cuts through the hills. All the precious land along the stream is covered with tobacco. Beginning in the winter months (November-December) tobacco takes the land through April, the perfect time to grow vegetables and subsidiary cereal crops.

All 15 Chak families in the 30 year-old Badujhiri Village are *jumias*. It has been a real jungle village for decades. Now it is rapidly losing its characteristics. The villagers can no more depend completely on *jum*. "The crops we get from *jum* are good for a maximum of six months a year. For the rest of the year we depend on the harvest of bamboo, which has also been exhausted," says Kijari Chak, *Karbari* of the village. "This situation has encouraged us to engage in tobacco cultivation to earn cash."

The Chaks have varied reactions to tobacco being brought to the area by the British American Tobacco. The company brings cash and chemical inputs (fertilizer and pesticides), the right incentives at a time when the villagers remain unemployed for months. "Tobacco has brought us employment for the months from November when the villagers have no work. Those with no grains in store can at least buy food from the markets," says Kijari.

The invasion of tobacco around Badurjhiri is a recent development starting in 2009. Twelve families got engaged in its cultivation in 2010. "Instant cash income is good from tobacco, which has expanded into the east from Badurjhiri," says A Thoai Ching Chak (30) of the village.

The Chaks of the Badurjhiri are awaiting a bigger threat that may come along with rubber cultivation. They have witnessed how rubber plantation on traditional *jum* land dispersed the villagers of their neighboring Chak village, Longoduhiri Chak Para.

Fearful of the invasion of rubber A Thoai Ching Chak says, "If rubber cultivation comes close to our village, our situation will be like the Chaks of Longodujhiri. We will be evicted."

It was in February 2011 when I last visited Badurjhiri Chak Para. Two Chak couples and a few other Chaks accompanied me to this forest village. We walked about four hours from Baishari Chak villages to reach Badurjhiri. I went back to Baishari in the last week of March 2013 and failed to convince the Chaks to take me to Badurjhiri village. They told me horrific stories of repeated attacks of the bandits on the Chaks there. The Chak families of Badurjhiri began to desert their village and taking shelter in Baishari Chak villages by early 2013. All women left Badurjhiri in fear of physical assault including rape. Many men have been physically assaulted. Cha Hla Khai (40) is one of them. We found him lying and languishing on a mat on a mud-floor in his make-shift house in Baishari. The bandits broke his right arm and bruised different parts of the body. The Chaks of Badurjhiri report that they can no more live in Badurjhiri that has been exposed to the outsiders. All Chak families are likely to abandon their homes in Badurjhiri [any time]! The Chaks believe that the key factor for the assault on them is expansion of rubber and tobacco cultivations around their village that has exposed them to the outsiders. The Chaks believe the bandits, many of them believed to be Rohingyas, want them to vacate the land. It will be no surprise if the village is totally encroached by the rubber cultivators and rubber trees are planted on the homesteads of the Chaks of Badurjhiri.

Rubber in Baishari—a story of deception: Rubber and also the so-called 'horticulture' have turned out to be a serious issue in four (Lama,

Naikhognchhari, Bandarban Sadar, and Alikadam) out eight upazilas of Bandarban Hill District. These have caused severe threat to the Chaks and other smaller ethnic communities and have caused massive destruction to the local ecology. These have brought in outsiders who have marauded the land that the indigenous communities have used for generations.

Land granted for rubber plantation and horticulture in Bandarban comes under the jurisdiction of three authorities—the Chittagong Hill Tracts Development Board (CHTDB), the Deputy Commissioner (DC), and the Standing Committee [for rubber]. The CHTDB oversees the rubber cultivation on 2,000 acres of land leased to 500 households under a rehabilitation project. These families are supposed to get land titles in their names, which they still have not gotten.

A much bigger area of land—around 45,000 acres—has been, according to a top official in the Bandarban Hill District administration, leased in 1,800 plots for commercial rubber production and horticulture. The size of an individual plot is 25 acres; in many instances more than one member in a family have got lease of land. The accurate records related to the status of rubber plantation and horticulture is very difficult to get. However, a government document provides a list of 1,635 individuals, proprietors, and companies who have received plots for rubber and horticulture in Lama, Naikhognchhari, Bandarban Sadar, and Alikadam upazilas. The leases were granted in a period from 1980 to 1996 (according to the government document). The list is an exposé of serious anomalies and a sinister government strategy. Of 1,635 individuals, proprietors, and companies that got plots of land for rubber and horticulture only 32 (7 in Lama, 9 in Naikhhongchhari, 14 in Bandaran Sadar, and 2 in Alikadam) are members of indigenous communities. The majority of the leaseholders are Bangalis— 210 from Bandarban Hill District (134 in Lama, 38 in Naikhhongchhari, 31 in Bandarban Sadar and 7 in Alikadam) and 1,393 from other districts [831 have plots in Lama, 340 in Naikhhongchhari, 131 in Bandaran Sadar, and 91 in Alikadam upazilas]. These individual owners include 174 proprietors and private companies [all owned by Bangalis].

One striking fact about these individual leaseholders is that they come from 45 districts outside the Chittagong Hill Tracts (CHT). The highest number of leaseholders (533) is from Chittagong, followed by Dhaka (367), Cox's Bazar (181), Comilla (66), Feni (35), Noakhali (34),

Lakshmipur (22), and Narayanganj (21).

The above picture of land sanctions for long term reveals further anomalies including grabbing of land not sanctioned. In the first place the land that has been sanctioned for rubber and horticulture belonged mostly to the category of "unclassed state forest", which to the indigenous communities is actually commons that they have used for centuries (in the Chittagong Hill Tracts more than 70% of land belongs to this category, while 24% is reserved forest and 1% is protected forest). Then the plots in most cases have not been properly demarcated. In general, land leased or possessed is marked by what is known as *chowhuddi*, which means the borders of land leased or in possession are determined by streams or trees or other objects at different directions. This provides scope to the leaseholders to take possession of much more land than what they have been sanctioned. That this has been happening in Baishari and elsewhere in Naikhhongchhari Upazila is widely alleged by the communities.

As recently as February 2011, forest patches in huge areas in 278 Baishari Mouza and 280 Alykhyong Mouza in Naikhongchhari Upazila were clear-cut for rubber plantation. A top official in the Deputy Commissioner's office, on condition of anonymity, said they know about the fresh land grabbing but are not able to take strong actions against the land grabbers.

The official said, "After the current government came to power we cancelled leases of 569 plots of rubber and horticulture on grounds that the lease holders violated the terms and conditions of the lease deeds. We cancelled some leases mistakenly. These include the ones actually planted with rubber. We cancelled the leases of plots to bring the land back from the grabbers. We will cancel more plots. The basis of cancellation has been our paper and the reports of the headmen. Around 35 plots cancelled have been recovered. Around 200 writ petitions have been filed with the High Court challenging our actions."

The DC's office finds it extremely difficult to bring back the land from the wrong hands! The well-wishers in the DC's office still hope to do good things to the indigenous peoples who are progressively losing control over their land. "The hill people know how to live with nature. We, the outsiders, have damaged their land and have caused them mischief. They must be given land ownership," said the official. "We want to see the interest of the local people protected first. In Bandarban Hill District half of 400,000 people are poor. We can offer two acres of land to each of the poor household with the land that we have got."

The DC's office confirmed that those who are purchasing plots from the primary lease holders include companies prominent among them being Destiny 2000 Ltd. and NGOs such as Quantum Foundation. These sometimes grab land in partnership with the local lease holders."

Although the DC's office and other government officials outwardly support the indigenous peoples, they do not see anything wrong with rubber plantation. "It is profitable; it is being exported from Bangladesh; and that the poor can also produce it is proven through the CHTDB project. But without ownership of land rubber plantation cannot be successful. We will soon give land titles to 300 participants to rubber cultivation under upland settlement project (second phase)," said the top official in the DC's office.

The chairman of the CHTDB, Bir Bahadur MP, also does not see anything wrong in rubber monoculture. "What is wrong is not doing it on the land leased for the purpose," says the chairman.

Mong Mong Chak (61), a retired high official in the CHTDB agrees with the CHTDB chairman (also an indigenous person) and the DC's office that see rubber as an investment in the national interest. "I am for and against rubber," says Mong Mong Chak who also puts blame on the Chak. "We, the Chaks, have lost control over land due to our foolishness. We did not apply for participation in rubber."

Kyaw Shwe Hla, the chairman of the Bandarban Hill District Council chairman has resentment about the way the land issues are being handled by the DC's office and the CHTDB. He resents the fact that his office does not have any information about rubber and horticulture plots and their current status. He makes mention of section 64 Ka of the Bandarban Hill District Council Act of 1989 according to which the council must give prior approval for lease, settlement, purchase, sale, and transfer of any land including *khas* land in Bandarban Hill District. "We want to know the current status of leases from the DC's office— how many have been cancelled, how many revoked, etc. The District Council was no way associated with all that has happened regarding leases of land for rubber and horticulture," said Kyaw Shwe Hla in February 2011.

One can imagine if the Bandarban Hill District Council does not have any statistics about the current status of leases and actions taken by the DC and CHTDB, how helpless the ordinary people at the community level can be. All that they know is the land grabbers somehow manage the upper people.

The Chaks, Marmas, and other local residents do not want rubber monoculture on the hills. "We are protesting against rubber cultivation in our area. Bir Bahadur MP, the CHTDB chairman came here and gave us his word that nobody would be evicted and everybody would stay where s/he is. But we do not see any sign of halting rubber. It continues to expand."

"Rubber has invaded the whole Baishari area. Our people are not aware. We receive allegations that rubber plantation is sometimes carried out in collusion with the headmen," says Bir Bahadur, MP.

Cho Hla Mong (42), the headman of 280 Alekhyong Mouza, a simple man on ground has a response to the allegation pointed to him by the CHTDB chairman. According to him rubber plantation has taken place in approximately 3,000 acres out of 16,808 in his *mouza* where only three indigenous persons including him have got rubber plots. The majority of the lease holders in his *mouza* come from outside and reside outside the locality (except for the Babu Company).

"I have given report in case of only two-three plots. Most of the plots have been leased without report of the headman," this statement of Cho Hla Mong testifies to his helplessness to the DC's administration and to the politicians who actually hold the power. He, of course, says that many rubber plots were sanctioned at a time between his father's death and his taking over the charge as headman.

The issue of rubber plantation is complex. It is profitable for those who do it right. Rubber and horticulture plots were given to the outsiders at the time of insurgency that came to an end in 1997 when a peace accord was signed. That the rubber plantation was one of many means to bring Bangali outsiders to the territory of the indigenous peoples with a purpose to outnumber them is quite obvious from statistics. This also demonstrates collusion among the business elites, politicians, local elites, and the bureaucracy. A legal framework was set in before initiating the rubber plantation that plays a role in transferring land to land-hungry non-indigenous people.

The 1971 Amendment to the Regulation 1 of 1900 Regulation,

better known as the Hill Tracts Manual, a key instrument for a separate administration for the CHT, for the first time gave the Deputy Commissioner the authority to settle land (generally up to five acres and in deserving cases 10 acres to a family—hill man or non-hill man) for rubber cultivation. "Settlement of land for rubber plantation exceeding 10 acres with a single family shall not be made without prior sanction of the Board of Revenue." The amendment also empowered the DC to settle land with an outsider with prior sanction of the Board of Revenue. The amendment of 1979 to the CHT Manual made a few significant changes to land settlement, especially for rubber plantation. This amendment awarded to the Deputy Commissioner such powers that he no longer required to take any prior sanction from the Board of Revenue for settling land [of any quantity] for rubber plantation and commercial purposes. The amendment (b-i) reads: "Land for rubber plantation and other commercial basis may be settled with a person on long term basis by the Deputy Commissioner up to 25 acres and by the Commissioner up to 100 acres. Settlement of land exceeding 100 acres shall not be made without the prior sanction of the Government."

The DC could freely exercise this power in leasing land to anyone till the signing of the peace accord in 1997 and leases of land till then remain valid. That's all the land grabbers need. Some come with papers in hand and use them to grab land that was never leased or settled. What concerns most is the pure land grabbers are roaming the area with axes in hand heading northeast, east, and southeast. They seem to be unstoppable.

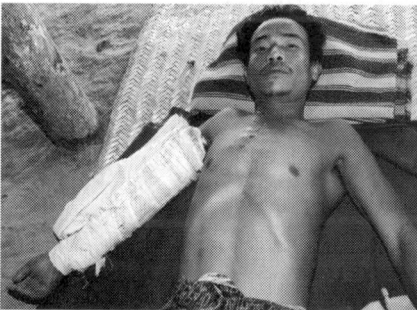

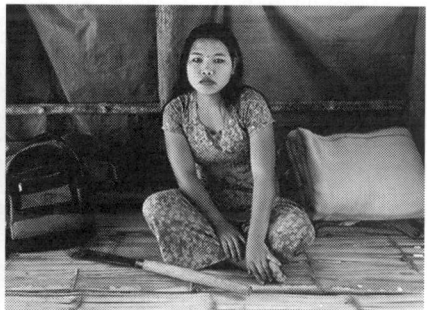

(Left) Cha Hla Khai (40), a Chak of Badurjhiri languishes in his makeshift house in Baishari. The bandits broke his right arm. (Right) A Chak girl of Badurjhiri in her temporary house in Baishari. All Chak women of Badurjhiri deserted their home in fear of attack and rape.   © Philip Gain

The CHT Forest and Its Demise

Historically what is Bangladesh now did not have any huge forest coverage. However, except for the Sundarbans, the largest contiguous mangroves on earth [located in the Southwest of Bangladesh], the natural vegetation throughout the whole of the Chittagong Hill Tracts used to be considered forests even few decades ago. But today the rich natural forest resources have become history.

The forests in the Chittagong Hill Tracts are broadly classified as being the tropical evergreen or semi-evergreen types, divided into four categories: (i) Reserved Forests (RF), (ii) Protected Forests (PF), (iii) Unclassed State Forests (USF), and (iv) Private Forests (PF).

Reserved Forest (RF): Before the advent of the British colonial administration and the creation of the reserved forest, the hill indigenous people had a free run of the entire CHT. They could freely cut timber and practice *jum* cultivation. But soon after taking control of the hills, the colonial authorities began to exploit the forests and in 1871 they declared almost the entire area a "Government Forest" (Schendel, Mey and Dewan 2000: 131).

The colonial authorities created the first Reserved Forest (Myani Headwaters Reserved Forests] in 1875 followed by Sitapahar Reserved Forest (1875), Matamuhari Reserved Forest (1880), Kassalong Reserved Forest (1881), Sangu Reserved Forest (1881), and Reingkhyong Reserved Forest (1882) (Forestal, Volume 5, 1964-1966: 2).

The CHT covers an area of 5,093 square miles (about 10 per cent of Bangladesh) of which the RF covers 796,160 acres or 1,244 square miles (about 24 per cent of the CHT) previously used by the *jumias*. The Forest Department directly administers the Reserved Forest. The four largest reserved forests today are Reingkyong RF, Kassalong RF, Sangu RF, and Matamuhri RF. There are a few smaller RFs that together cover 24 square miles (Webb and Roberts, 1976: 2 in Halim, Roy, Chakma, and Tanchangya).

Initially the reservation was intended "mainly toward preservation of the forest from juming and illicit felling and toward creation of plantations designed to gradually replace the less valuable species with more desirable species." (Forestal Volume 5, 1964-1966: 2). What is clear from this British colonial intention of reservation is that the local species and biodiversity of the CHT was not valued, which created the grounds of colossal damage to the native forests and the biodiversity it sustained.

Subsequently, the government of East Pakistan started planning a full-scale industrial exploitation of forest resources and established an autonomous body, Forest Industries Development Corporation (FIDC) [by the Ordinance No. 67 of 1959]. After independence this body was renamed, Bangladesh Forest Industries Development Corporation (BFIDC). The government plans for extraction of timber through FIDC and later BFIDC from the natural forests of the Chittagong Hill Tracts has had devastating effects on the native forest and the unique ecology of the CHT. The corporation began timber extraction from Karnafuli Valley Reserve Forests in 1960 and from Sangu Matamuhuri Reserve Forests in 1963 (Khan in Banglapedia 2003: 481). The corporation has, to date, extracted timber from 70,729 acres of forestland in the reserve forests of Kassalong, Reingkhyong, and Sangu Matamuhuri and handed the vacant land to the Forest Department for raising "dense forests" (Khan in Banglapedia: 2003: 482). The creation and operation of this BFIDC demonstrates the state attitude towards the forest and its devastating effects on ecology. What is "dense forests" to the state is nothing but monoculture of teak and a few other species. Most of the reserved forests are cleared of their natural stands today. The areas that once had natural stands of gigantic trees are barren or have some coverage of plantations with no wildlife and the diverse plant species that support such wildlife. The topsoil in the areas without vegetation or with [teak] plantation is exposed to rains and erodes quickly. This also causes landslides. The loose soil is washed down into the Kaptai Lake.

In addition to the reserved forests from the colonial times, there has been further expansion of such forest by notifications under the Forest Act. According to information of the Committee [now Movement] for the Protection of Forests and Land Rights in the Chittagong Hill Tracts [formed on September 1, 1998], 217,790.3 acres of land from 83 *mouzas* in three hill districts have been notified reserved forests between 1990 and 1998. Of these 140,341.31 acres have been finally notified as reserved forests. However, a BBC Radio broadcast quoted a high official in the Ministry of Environment and Forest (MoEF) saying that 208,148 acres had been primarily notified as reserved forest of which 116,880 acres had been finally notified.

The hill people complain that the government has arbitrarily reserved the *jum* lands, croplands, orchards, homesteads, and land registered in the Deputy Commissioner's office.

According to the estimate of the Committee (now movement) for the Protection of Forests and Land Rights in the Chittagong Hill Tracts, if the government plan for the expansion of reserved forest from 1990 is implemented, around 200,000 people will be affected.

The government has the law in hand. So it has declared the forestland used by the local communities as reserved. The local communities consider the expansion of the reserved forests as an immoral act. It dispossesses the local communities of their livelihood means, and causes immense harm to biodiversity and to the knowledge about medicinal plants of the local indigenous communities (Gain 2000: 25). Expansion of reserved forests, mainly for plantations backed by the IFIs and donors, has further complicated the land and ecological questions in the CHT.

Reservation Hits the Khyang Most

Sangthuima (24) and Thuisangma (20), two Khyang sisters in a remote village in the Chittagong Hill Tracts (CHT) have had almost all of their three-acre croplands notified as reserved forests. Their homesteads have also been notified as reserved forests. They have implicitly become illegal on their own land where they and their forefathers have lived for centuries. The two sisters inherited the land from their father, Teng Hla Pru and the land is still recorded in his name.

The others of the same hamlet—Baro Kukkachhari Para in 335 Dhanuchhari Mouza of No.1 Gilachhari Union in Rajasthali Thana in Rangamati Hill District, face the same fate. Their croplands and homesteads have been notified reserved.

One obvious impact of the reserved forest expansion is the disappearance of many Khyang families from the area. The simple, quiet and peace loving Khyang people, who numbered 2,343 in 1991 (population census report), find it difficult to cope with the changed situation as a result of reserved forest expansion up to their garden land and homesteads.

The headman of Kukkachhari Mouza (in Rajasthali Thana), Mishu Marma complained that the size of her *mouza* is 666 acres; but officially 722 acres have been reserved in 1984/85 and trees such as Acacia, Gamar, Kadam [under plantation scheme to produce raw materials for the Karnaphuli Paper Mill (KPM) in Chondroghona in the CHT region] have been planted.

"In our *mouza* we all have become illegal residents. Out of 30 Khyang families in my *mouza*, 23 have disappeared and of 600 families in my entire *mouza*, 200 have virtually become refugees. The others still live here because we still cultivate some lands and live on our homesteads, which have also been declared reserved forests," said Mishu on October 17, 1998.

In another village, Arachhari Headman Para (in 335 Dhanuchhari Mouza), the headman of the *mouza*, Hlathwai Khyang informed that all of the 33 families in the village are Khyang. Previously there were about 60 Khyang families in the village. He complained that families have disappeared because of harassment and forest cases especially after the expansion of the reserved forests. According to his information, of 3,969.98 acres in his *mouza*, 3,889.31 acres are reserved forests and 139.72 acres are registered land, which include 55 acres registered to 11 families (forest villagers)—five acres each.

Headman Hlathwai Khyang and others informed that Dhanuchhari Para, Korbanchhari Para, New Zealand Para, Madan Karbari Para, Moniong Karbari Para, and Bara Kukkachhari Para of Dhanuchhari Mouza are today inhabited by only 180 families and most of them are Khyang. But beforehand there were many hundreds more Khyang families in these hamlets.

Why and where have all these families gone? "The plantation activities and conversion of the *jum* land, garden land, and homesteads into reserved forests have chased them out of their traditional homeland and pushed them upward where they can look forward to practising their traditional economic activities for a living," says Arun Laibresaw, a Khyang in Arachhari Headman Para.

Headman Hlathwai Khyang said that in his *mouza* the local people have applied for registration of around 700 acres of land in the DC's Office. The settlement of those lands, classified by the Forest Department as Unclassed State Forest (USF), has been stopped since 1989. This is alarming for the Khyang.

Expansion of reserved forests by notifications and restriction on *jum* cultivation has made life extremely difficult for the communities in the remote areas. "The Khyang people are most affected by the plantation activities of the Forest Department (FD). Almost the entire Khyang population of the CHT has been affected," says Raja Devasish Roy, the Chief of the Chakma.

In Dhanuchhari Mouza the Khyang can no more practise *jum*. Without *jum*, the subsistence economy of the Khyang people is totally ruined. "We used to produce all our food items in *jum* fields. Since 1976 *jum* cultivation has been stopped because of the expansion of pulpwood plantation and reserved forests. We can now produce little food. We have to buy foods, which are expensive for us and we have to walk longer distances to buy them. We are now in a tough struggle for survival," says Aung Saw Khyang of Arachhari Headman Para.

An example how the expansion of reserved forests has affected the registered land is the case of Gyan Bikash Tangchangya of Rajasthali Thana Sadar. He complained that the Pulpwood Plantation Division of Kaptai forcibly planted pulpwood trees on five acres of his land (in Rajasthali Thana) in 1980s. He registered this land in 334 Kukkachhari Mouza of Ghilachhari Union in Rajasthali Thana in mid-1980s. The division had forcibly planted pulpwood on many others' land, complained Gyan Bikash.

Advocate Dinonath Tangchangya, Gyan Bikash's son complained, "In 1995 officials from the division began harvesting trees they had planted without consulting the local people who assert claims on the land."

Probhat Kumar Tanchangya, son of Minunath Tanchangya registered five acres of land in 335 Dhanuchhari Mouza in Rajasthali Thana in the 1980s in which he planted timber trees—teak and gamar. Now after many years of hard labor Probhat's trees have begun to mature and he wants to harvest some timber. Recently, he had almost completed the process of procuring *jote permit* for harvesting 948 cft teak and 500 cft fuelwood with the approval of the district forest committee. The DC's office had requested the DFO (Pulpwood) of Rangamati in a letter dated June 6, 1997 to issue a *jote permit* to Probhat Kumar Tanchangya. To his great disappointment, the DFO refused to issue the much desired permission document and told him that his land was reserved forest then.

A Khyang village in Rajasthali Thana in Rangmati Hill District.
© Philip Gain

Protected Forests (PF): The PF covers 34,688 acres or 54.20 square miles (about one per cent of the CHT). Most of PF has recently been categorized as reserved forest (Roy 2002: 16). While the District collectorate administers the PF, the Forest Department manages and controls its forest resouces.

Unclassed State Forest (USF): The rest of the CHT, USF covers 2,463,000 acres or 3,848 square miles (about 75 per cent of the CHT). The people of the CHT used to freely collect timber and bamboo and practice *juming*. The timber and bamboo that were in abundance in the past are now very limited. *Jum* cultivation is still practised but much of the land previously available for *juming* is leased or encroached upon for rubber cultivation, pulpwood plantation, settlement, tree gardening, etc. to the disadvantage of the hill indigenous peoples. The District Collectorate and the District Councils, in conjunction with the *mouza* headmen control USF (Roy 2007: 62).

Some Key Features of the CHT Forests: The flora of the CHT forests is distinctive in character and resembles the flora of the Arakan. However, the teak patches that we see throughout the CHT are planted forests, not indigenous to the CHT.

The most important commercial timber species of the Chittagong Hill Tracts used to be *Jarul, Gamar, Garjan, Chapalish, Toon, Koroi, Civit, Champa, Simul, Chandul,* etc. These trees, among 100 recognized tree species grow to gigantic heights. Added to these are numerous undergrowth shrubs and brush-like species (Forestal, Volume 5, 1964-1966: 3). Most of the trees in the undergrowth are of the evergreen type and most of the tallest trees are deciduous and semi-deciduous. Some of the trees shed their leaves during the cold season and some in the summer. So the forests always look green, or more correctly, the forests never lose their semi-evergreen picture. However, this is a description of the forests that were. Most of these trees of the native forest have disappeared.

Apart from evergreen and deciduous forests, bamboo, and savannah types are of immense economic and environmental value. Bamboo grows as 'undergrowth' under various types of forest. The most important industrial use of bamboo is as raw material at the Karnaphuli Paper Mill (KPM). It is also used for house construction and it supports many cottage industries. Bamboo shoots are important food for the hill people. The significance of bamboo was glowingly illustrated by Captain Lewin, one of the most sympathetic and knowledgeable of the British Chittagong Hill Tracts' Commissioners. (See Sudibya Kanti Khisha, P 131, for Lewin's description of bamboo).

The stock of bamboo in the hills is drastically reduced nowadays. Millions of tons of bamboo and softwood have been cut to feed the KPM at

Forest

The Chittagong Hill Tracts (CHT),
Chittagong & Cox's Bazar districts

Bangladesh 2006

Index map
Bangladesh

Area of ...

Legend

Mapping Unit	Main Land use
	Mixed evergreen & deciduous forest (including reserved forest)
	Mixed thickets & forest
	Planted mangorve forest
	Reserved forest

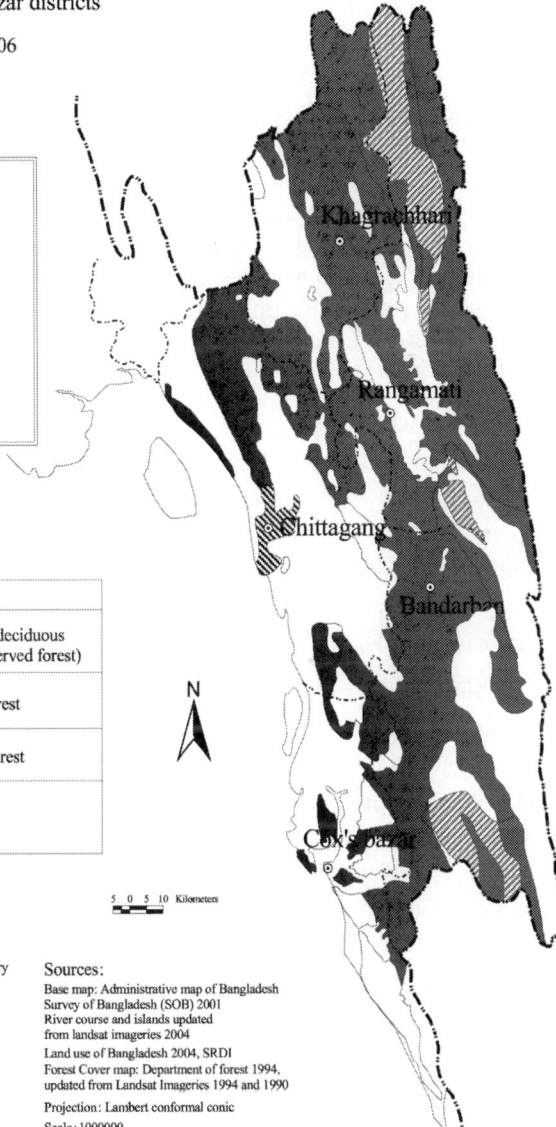

N

Conventional signs

	International boundary
	District boundary
	Narrow river
⊚	Head Quarter
	City
	Kaptai lake

5 0 5 10 Kilometers

Sources:
Base map: Administrative map of Bangladesh
Survey of Bangladesh (SOB) 2001
River course and islands updated
from landsat imageries 2004
Land use of Bangladesh 2004, SRDI
Forest Cover map: Department of forest 1994,
updated from Landsat Imageries 1994 and 1990

Projection: Lambert conformal conic
Scale: 1000000

Chandraghona set up in 1953. The promotion of plantation economy has also contributed to the drastic loss of bamboo stocks.

The savannah type of forest is where there are practically no trees and the areas are invaded by sungrass. A very large portion of the USF comes under this type of forests.

It is not only the trees and mountains that dominate the CHT, its fauna also makes the area different. The wild and mighty elephant is one particular attraction of the CHT. Once bison, sambur, barking deer, leopard, Bengal Tiger, panther, and other animals were found all over the CHT. Most of these animals have disappeared or have drastically decreased in number today. The tiger species has become totally extinct in the CHT jungle. Attractive CHT bird species include Imperial Green pigeon, White-winged Wood-duck, Hill Mynah, Greater Racket-Tailed Drongo, etc.

Forest Land in the CHT

Classification	Rangamati (ha)	Bandarban (ha)	Khagrachari (ha)	Total CHT (ha)
Gazetted reserved and protected forest (RF+PF)	234,520	74,841	23,151	332,510
Estimated remaining reserved forest (FF)	49,613	-	4,018	53,373
Encroached RF*	2,176	-	-	2,176
Estimated remaining protected forest PF	0	0	0	0
Planted forests (private)	**22,259**	**26,184**	**8,930**	**57,373**
Unclassified state forest (USF)	**322,521**	**292,522**	**94,656**	**709,699**
USF notified for reclassification to RF	23,680	27,000	12,660	63,340
TOTAL	**579,300**	**393,547**	**126,737**	**1,099,584**
Total in district under control of FD (thought 'notified' land, not formally 'gazetted')	258,200	101,841	35,811	395,852
Forest area controlled by Ministry of Land	322,521	292,522	94,656	709,699

*Similar activities occur in Bandarban and Khagrachari on non-RF land

Soruce: FD (Dhaka) as cited in ADB (2001b, p 18) in Raja Devasish Roy 2002 .

Forest area by forest types (1000ha)

Hill forest:	Sal Forest:	Mangrove	Bamboo
551	34	Forest: 436	Forest: 184
Long Rotation	Short Rotation	Mangrove	Rubber
Plantation: 131	Plantation: 54	Plantation: 45	Plantation: 8

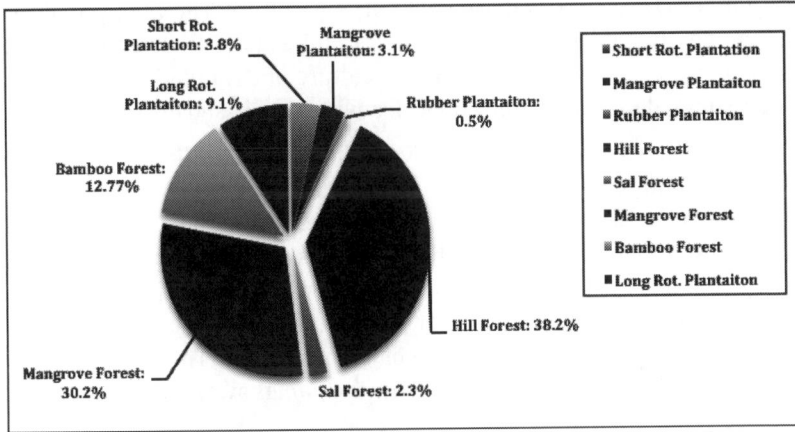

Private Forest (PF): Forest [plantations in essence] that the indigenous farmers own. Non-resident outsiders (individuals and companies) also own a few plantations.

There are nine forest divisions in the CHT—six in Rangamati Circle CHT: South Division, CHT North Division, Jum Control Division, USF Division, Kaptai Pulpwood Division, and Khagrachhari Forest Division and three in Chittagong Circle: Bandarban Forest Division, Lama Forest Division, and Bandarban Pulpwood Division.

Jum and *Jumia*

Swidden *or Jum* (shifting) cultivation, of the hill indigenous peoples, children of forests, has been a predominant land use system in the CHT for centuries despite the government efforts to stop it since the British period. Till 1818, it was the only form of agricultural practice by the hill people in the CHT. By 1960 the land area under *jum* was 41,885 ha with fallow period of 8 to 10 years. (*Jum* used to be a practice in some other regions by some indigenous communities. Modhupur *sal* forest is one such area where the Garos practised *jum* till 1950 when the government banned it).

The farmers possess titles for their land in the valleys and they can transfer or sell this valley land. But the land they use for *jum* agriculture belongs to the state and they have no individual rights on *jum* plots that they prepare in unclassed state forest (USF). They have rights only on the crops grown in *jum*.

Jum cultivation is a kind of community agriculture. The whole family or community might get engaged in it that requires hard labor and constant care. It also has great influence on the culture of the indigenous peoples. Their songs, dances and many other festivities are based on *jum*.

Jum is carried out predominantly on the steep slopes of the high hills. A *jum* cultivating family or community (*jumia*) selects a convenient piece of land during the months of January and February. The *jumias* then cut all vegetation and trees except the large ones. The cut *jum* is then dried in the sun. It is set on fire before the monsoon rain begins (in April-May). Thoroughly dried vegetation cover is reduced to ashes by burning. The fire might burn the soil to the depth of an inch or two. The ground is then cleared of charred logs and debris. Now the *jumias* await rain.

As soon as rain falls and saturates the ground, sowing commences. The *jumias* mix seeds of different crops—rice, cotton, melon, pumpkin, millet,

Jumia at harvest time.   © Philip Gain

beans, gourds, yam, maize, oil seeds, ginger, sesame, etc. —and plant them in small holes at fairly even intervals. The different crops ripen at different times; the maize ripens at about the middle of July; melons, vegetables and rice are harvested in September and October; and cotton and sesame in November and December. Bananas, if planted, are left for harvesting in the second year.

After harvest the *jum* fields are left to themselves. The undergrowth and bamboo soon cover the soil.

A successful *jum* cultivation gives considerable results. A *jum* can be cultivated only two to three years in a row and then the land takes approximately 10 years or more to recover. In recent times [50 to 60 years], the fallow period has become dangerously shortened from its traditional 10-year fallow period to 2 to 3 years. However, according to experiments in many countries throughout the globe even the 10-year fallow period is not long enough for shifting cultivation to be sustainable. Geoffrey Scott who studied the situation in the Peruvian Amazon has concluded that shifting agriculture appears to be sustainable if the fallow period is 15 years or more (Marchak 1997:165).

The village rather than individual cultivators usually manage *jum* land. The headmen distribute *jum* land among the village community. The *jumia* families pay a tax to the headmen to be shared by the headmen, the chiefs and the government. "*Jum* land 'belongs' to the one who occupies it first, and registers with the headman." In this context the state is the ultimate owner of all land in the Chittagong Hill Tracts (CHT Manual, in *CHT District Gazetteer*). The Deputy Commissioner is empowered to control and regulate *juming*.

The government and the international agencies involved have consistently regarded shifting cultivation as a bad agricultural practice. The governments from the colonial times and these agencies [Forestal Forestry and International Engineering Limited and Asian Development Bank among others] have blamed *jum* cultivation for deforestation and degradation of environment. So they have always tried to control and discourage *jum* cultivation. But the state and the supranationals, which have interest in the CHT, have hardly given pragmatic consideration to the factors that have hastened the competition for land and severely shortened the fallow period. The scientific analysis of *jum* cultivation has also not been accepted by the policy planners.

The British colonists and subsequent governments have consistently condemned the *jum* cultivation for destruction of forests. But they have downplayed factors such as infrastructure development (not necessarily in the interest of the local communities), Bangali settlement, ill effects of militarization and human rights abuses, industrial plantation, expansion of reserved forests into the *jum* land, leasing *jum* land to the non-locals for commercial purposes, logging, etc.

The British government introduced the Forest Conservancy Rules in 1876 to limit *jum* cultivation (Ishaq 1975). This encouraged bringing plough cultivators into the hills from the plains.

Land use policy changes in 1970s and in 1980s, allowed leases of the CHT land to non-residents and corporate bodies, especially for establishing industries and raising commercial plantations. In most of the cases, these lands remained totally unused and many of these lands were and in some cases still are occupied by indigenous people who have been living on and cultivating them for generations. During the same period the Jum Control Division of the Forest Department and the Chittagong Hill Tracts Development Board (CHTDB) undertook programs to rehabilitate *jumias* through horticulture and agroforestry. Such attempts failed and an overwhelming majority (80% according to an estimate) of the settled families deserted the settlement villages because they did not have adequate means to survive under the government plan.

In the recent times, government forest policies, the forestry master plan, different forestry projects, amendments to the 1900 Regulation, amendments to the Forest Act of 1927 in 2000 have serious negative effects on the *jum* cultivators. In many places throughout the hills, the *jumias* have to give up *jum* cultivation and become day laborers in the government-sponsored plantations.

In the discussion on *jum* one needs to understand the essence of this traditional agricultural practice. It is a traditionally developed agricultural practice, which is fit for the mountain regions throughout Asia and the globe. *Jum* cultivation takes over 30% of the world's exploitable soil of which more than 40% are in Asian countries. In Northeastern Indian Himalayas, *jum* is a predominant land use system supporting 1.6 million hill people over an area of 426 million ha. In the Nagaland State of the Indian Himalayas, an estimated 1,000 sq.km of sloping land was brought under shifting cultivation in just one decade and this increase was partly due to reduced fallow period from 14 years to 5 years (Partap 1998).

Given that the land selected for *jum* cultivation is burnt, this agricultural practice is commonly blamed for deforestation and soil degradation. During the preparation of *jum* every year, we see photos of burning vegetation in the CHT and the *jumias* are branded to be the despoilers. This is certainly a misrepresentation. All the trees are not felled during the preparation for *jum*. A good number of tree species are kept in *jum* field. This helps maintain biodiversity (Alam and Khisa 1998). Experts reject a common belief that *jum* causes loss in soil fertility. The scientists who have studied the effects of shifting cultivation in Latin America (especially Brazil) between 1974 and 1983, conclude, "Nutrient stocks in the soil did not decline greatly during the period of cultivation, but there was a decline in phosphorus, which may have been responsible for crop-productivity decline. Following abandonment, the productivity of native vegetation increased rapidly. The native plants were apparently better able to adapt and acquire phosphorus than crop plants" (Marchak 1995:164).

In the Chittagong Hill Tracts the governments assisted by the international financial institutions (IFIs) and donors have established teak, rubber and pulpwood plantations. Such monoculture plantations cannot be alternatives to *jum* cultivation, which has worked well and without too much detriment to the soil fertility.

There are other interesting findings about *jum*. One is that *jum* cultivators change the land, not the variety of crop. This protects their land and crops from diseases, insect attacks and weeds (Whitmore 1998:157). A successful *jum* also gives considerable results.

Given the limited economic choices that the hill people have and the geographic conditions, this centuries old agricultural practice will move on, especially in the remote hill areas. R.H. Sneyd Hutchinson was perhaps correct as he observed that the natural feature of the CHT was such that *jum* culture must be the principal method of cultivation and he argued that if the system of *juming* was abandoned, the hill ranges would be entirely useless. However, conditions created because of land use policies taken during the British rule and afterwards pose unprecedented threats to *jum*, *jumias,* and the local environment.

Monoculture Plantations

Monoculture plantation, "simple" in most part with exotic species has been the single greatest threat to the native forests in the hills [as well as in the plains] thus affecting the life and livelihood of the forest dependent communities. Plantations in Bangladesh include teak, pine, pulpwood, rubber, agroforestry, woodlot (for the production of fuelwood), etc. The Forest Department and Bangladesh Forest Industries Development Corporation (BFIDC) carry out these plantations on the public forestland, most of them in the Chittagong Hill Tracts (CHT), Chittagong, Cox's Bazar, Sylhet and the *sal* forests. Most of the recent plantations have been funded by international financial institutions (IFIs)—ADB and World Bank in particular. These recent plantations in Bangladesh are 'simple' and are the consequences of the investment strategies of these two IFIs and external influences.

Back in the eighties in the last century, the government policy on plantation was: "To re-place the existing irregular, depleted and less productive forests by a man made plantation forest with more valuable and productive species suited to the soil and country's requirements" [(*Revised working plan for the forests of Chittagong Division*) for the years 1978-79 to 1987-88]. This government policy has had far-flung negative consequences on the native forests and the forest dependent communities. The major monoculture plantations in the CHT are:

Teak: The British colonial foresters introduced teak (*Tectona grandis*) to the Chittagong Hill Tracts in 1872 with seeds from Burma (Myanmar). Teak was considered an important species for the future plantation

Teak plantation in the CHT, a big factor for ecological damage. © Philip Gain

program. In 1947-48, teak was planted 334 ha and in 1952-53, another 1,194 ha plantation was raised mostly with teak. The Forest Department in Bangladesh has continued the teak plantation ever since. The Pakistan and Bangladesh governments followed the colonial policy of teak monoculture. In gardens raised by individuals, teak is the most desired species. Teak produces desired timber, but does not necessarily help the local economy. It is also damaging to the local environment. A handful of individuals benefit from teak plantation.

The understory vegetation is absent in the teak plantation because of an acidic reaction with the soil. Teak is certainly not a replacement for any local species. "A monoculture of teak, however desirable commercially, is a death-knell to wildlife conservation. No species of deer or monkey can find food in such vegetation, and they are forced to abandon the area. In the absence of deer, tigers and leopards turn to preying on cattle and goats around the villages and are therefore shot ..." (van Schendel et. al 2000:147). Even this description of the teak plantation is outdated. Tigers and leopards have gone extinct long ago in the CHT.

Pulpwood: Pulpwood plantation, also monoculture, is a major environmental and social concern in the CHT. There are two pulpwood divisions in the Chittagong Hill Tracts (CHT). One is the Pulpwood Plantation Division Kaptai in Rangamati Hill District, established in 1978 by and executive order on 144,546 acres of land. The other is the Bandarban Pulpwood Division established in 1982. Initially 119,321.60 acres of land were assigned as work area for the division; later on in 1986 another 54,962.80 acres were added to it. However, only 33,088.75 acres (including 17,224 reserved forest) of land were handed over to the Bandarban Pulpwood Forest Division by 2012.

The agricultural practice of the *jumias* was blamed to establish the rationale for pulpwood plantation. "A vast area of Un-classed state forests (about 22.38 lakh acres) were in critical conditions due to repeated jhooming of indigenous people as their traditional practices. The traditional shifting cultivation in the USF areas resulted in severe soil erosions, degradation of soil fertility, silting-up of main rivers along with their tributaries causing floods both inside as well as outside of CHT & also causing threats to Hydro-electric project," writes a forest official (anonymous). The forest official refers to the survey of the Forestal Forestry & Engineering International Ltd. Canada during 1964-66. The experts engaged to conduct the survey, according to the forest official, recommended to put 3.5% of the land under agricultural practices, 20% under horticultural practices and the rest 76.5% under permanent tree cover.

The Swedish International Development Authority (SIDA) and Bangladesh government shared costs (fifty-fifty basis) of Taka 190 lakhs of the first pulpwood plantation project, "Development of pulpwood plantation in the CHT forests" implemented by FD during 1976-77 to 1979-80.

The argument for pulpwood plantations is fallacious. One clear object of the pulpwood plantation in the CHT region is to replenish the shortages of natural pulpwood for the Karnaphuli Paper Mill (KPM) at Chandraghona. At one time it was thought that the raw materials for the KPM would never be exhausted. This turned out to be a myth. Since it went into operation in 1953 the gigantic mill has consumed millions of tons of bamboo and softwood that came from the forest of the CHT. And the natural stocks of raw materials have drastically reduced. It is in this backdrop that the pulpwood divisions were created leaving tens of thousands of ethnic people of the CHT at a grave loss.

The pulpwood plantations established mainly on the USF that is also the traditional *jum* land have caused loss of biodiversity and soil erosion. Insect attack and diseases in these monoculture plantations are also reported.

Rubber monoculture—a green desert: Rubber plantation has replaced significant portions of the forest patches in the Chittagong Hill Tracts [other areas of rubber plantations are Modhupur *sal* forest and Sylhet in particular]. These are basically areas inhabited by the *jumias* and Adivasis. The government agency that has raised rubber plantation is Bangladesh Forest Industries Development Corporation (BFIDC) that has raised 32,635 acres of rubber plantation [2005]. Of these 16,617 acres belong to the Chittagong zone, 8,083 acres to Sylhet and 7,935 to Tangail (in the Modhupur *sal* forest).

As has been explained above rubber and horticulture on 45,000 acres of land of which 32,500 acres for rubber (Hossain 2013) has caused a horrific situation for the Chaks and other smaller ethnic communities in four upazilas in Bandarban Hill District.

The rubber plantation in Khagrachhari to rehabilitate *jumia* families under an Upland Settlement Scheme (USS), a project financed by the Asian Development Bank (ADB) also demonstrates how the forest-dwelling indigenous peoples have been deceived. Starting in 1979, 2,000 indigenous households were involved in the first phase. Each household got 4 acres of land for rubber, two acres for horticulture and .25 acres for homestead. In the second phase starting in 1993, 4,000 acres of land was allocated to 1,000 families (4 acres to each) for rubber cultivation. In this phase each family got 1.25 acres for horticulture and homestead. So far, 500 families have got settlement for horticulture and homesteads (2.25 acres each). None of families has yet got settlement for rubber plot.

Rubber, be it in the CHT or Modhupur, may bring some benefits for BFIDC and the private entrepreneurs, but in general it has not been beneficial to the people who once used these land as commons.

Rubber plantation is a clear threat to forest and genetic resources. It is a single species monoculture that takes places on the forestland. All species have to be wiped out from an area demarcated for rubber plantation. This is a clear ecocide no matter what financial benefit is accrued.

A rubber plantation may look like a forest from a distance but it is a desert for birds, insects and other animals. Rubber is a mono-crop that grows in beautiful rows but does not support any life forms. It is a green desert indeed.

The government had a policy to discourage rubber (starting in 1959

Rubber monoculture on forestland in Bandarban—profitable for oursiders but it causes wholesale damage to local ecology.   © Philip Gain

as an experiment) till the end of 1970s when at the recommendation of the CHT feasibility study team undertaken by UNDP and ADB [in 1978], rubber was included as a major crop in the upland settlement component of the CHT multi-sectoral development program. One key motivation for rubber plantation in the CHT was alleged to be to counter insurgency.

Tobacco: The Government statistics show that tobacco was cultivated on 79,415 ha of land in 2010, which according to Department of Agriculture Extension, shrunk to 52,922 ha in 2011. However, according to local farmers, elected chairmen, and members of Union Councils, field workers, agriculture officials and local agents of tobacco companies tobacco was cultivated on 96,000 ha, which came down to 79,500 ha in 2011 (*Prothom Alo* 31 May 2011). Traditionally Rangpur and Kusthia have been two districts with most of the tobacco cultivated in the country, while it is cultivated in all other districts in small quantities. While these two districts still produce a large percentage of tobacco of the country, its cultivation kicked off in the Chittagong Hill Tracts after 1990 and has expanded ever since. Among the three hill districts Bandarban Hill District produced the highest quantity of tobacco in 2009-2010.

Tobacco cultivation that has expanded rapidly in Bandarban in particular is seen to be a serious threat to traditional agriculture, and the food security of the local communities, and environment. It is cultivated on precious flat land that the hill peoples used for production of vegetables and other cash crops particularly for their own consumption. The invasion of tobacco in the remote areas also brings chemical inputs (fertilizers and pesticides) that pollute streams, source of potable water that is also used for other purposes.

A key environmental concern with tobacco cultivation in the CHT, Chittagong and a few other districts in the west of the country is the burning of wood to dry the tobacco leaf. According to a report 30% of the forest destroyed in Bangladesh is destroyed due to tobacco cultivation (*Prothom Alo*, May 31, 2011). Tobacco cultivators in Bangladesh dry their green leaf either in the sun or in *tondur*. *Tondur* is a kind of sealed chamber or house with mud-walls. A *tondur*, set up for one ha of land normally requires 500 *maunds* (a maund is approximately 40 kg) of firewood.

Momtajul Haque from Cox's Bazar who has a two-acre tobacco plot in Longodujhiri (Khal) Chakpara, a remote village in Naikhhongchhari of Bandarban Hill District [now abandoned], said in 2011 that he required 500 trees to dry his tobacco leaf. All these trees came from the nearby jungle. According to a newspaper report tobacco cultivation engaged

12,274 ha (30,329.7 acres) of land in three districts of the Chittagong Hill Tracts, Chittagong, Cox's Bazar, Meherpur, Jhenaidah, Chuadanga, Rajbari and Chalan Beel area. According to the local farmers, elected chairmen and members of Union Councils, field workers, agriculture officials, and local agents of tobacco companies, in these nine districts tobacco was cultivated on 26,388 ha (65,206 acres) of land. If 500 maunds of wood is burned for drying tobacco leaf of one ha, an astounding amount of wood (13,194,000 maunds) is burned for drying tobacco grown in this vast land. (Abul Hasnat, *Prothom Alo*, 31 May 2011).

Health Risk from Tobacco: Nicotine, which can cause cancer is the main addictive element of tobacco. It is neurotoxin and works on Amitailcolin Receptor. There are some other substances in tobacco that can cause cancer. Oral ingestion of tobacco that is used for making cigarette can cause death. Heart problem, breathing problem, gastric, back pain, headache, fever, asthma, skin problem (dry and yellow skin), coughing, vomiting, dizziness are some major health risk found among tobacco cultivators. "Barzus" and "Green Tobacco Sickness" are mostly found among the family members engaged in tobacco farming (Arnab 2011).

Smokeless tobacco use including *jarda, sada pata, khoinee* and *gul* also cause health problems like cancer of lungs, mouth, larynx, pharynx, esophagus, kidney, pancreas and bladder.

The use of tobacco is more fatal for the pregnant women. Among pregnant women it can cause an increased risk of giving birth to under-sized infants, preterm delivery, miscarriage, neonatal death, low birth weight (among fetus) and others. Children born to women who smoke are more at risk of cold, respiratory problems and ear-aches.

Poppy and *Ganja*: The production of poppy and *ganja* (cannabis herb) is illegal in Bangladesh and punishable offense up to death as provided by the Narcotics Control Act, 1990. Yet these are grown in some districts including the Chittagong Hill Tracts (CHT).

According to a press report 15,000 acres of land were used for poppy and *ganja* farming in the Chittagong Hill Tracts in 2003 with a production worth Tk.500 millions (*Prothom Alo* 2003). These are grown in the remote and inaccessible areas in the hills between September and February (right after *jum* season). These are grown also with other *jum* crops. At the end of the November, farmers plant seeds; by March-April poppy plants become mature to produce opium and heroin. And it is the right time to collect liquid from a seed of the poppy flower that is the main ingredient of heroin and opium. Other drugs produced from purified opium include Morphine, Codeine, and Papaverine. Heroin comes from Morphine (Chakma 2003).

An insurgent group of Myanmar "Arakan Liberation Party (ALP)", drug smugglers, local *headmen*, *Karbaris* (traders) and *Bazaar Chowdhary* (market caretaker) reportedly provide necessary raw materials (money, seeds, insecticides, etc.) to the local hill people for growing poppy and *ganja*.

The cultivation of poppy and *ganja* requires less labor and is more profitable compared to other crops grown in *jum*. Better pay to laborers in the poppy gardens than other *jum* work is an incentive for opting for illegally growing of narcotic plants.

Concentration of poppy and *ganja* production in the hills include Kutukchhari and Sapchhari unions of Rangamati sadar upazila; Kaptai, Naniarchar, and Kawkhali upazilas in Rangamati; Matiranga, Laxmichhari, Ramgarh, Manikchhari, Mohalchhari, Panchhari, and Dighinala upazilas in Khagrachhari; Bandarban sadar, Thanchi, Ruma, Lama, and Alikadam in Bandarban. Poppy and *ganja* in these areas are grown along the banks of the rivers, streams, and in the forests. The cultivators cut the trees in the deep of the forests to expand the cultivation of these illegal crops that has reportedly caused massive destruction to the Sangu Reserve Forest of Thanchi upazila in Bandarban (*Prothom Alo*, February 27, 2005).

A special unit of Bangladesh Army is assigned to regularly conduct raids against the illegal narcotic plants in the hills. In a recent raid (2010) the army reportedly destroyed 122 poppy gardens in Bandarban Hill District (*The Daily Star* February 25, 2011). Bangladesh Army with support of BDR (now Border Guard Bangladesh-BGB), Narcotics Control Department, and the police also conducted raids several times as well. Besides, the Army with the help of local inhabitants, leaders, and administration has initiated awareness programs against farming and use of such illegal crops among the local farmers. Incentives in vegetable seeds, fertilizers, insecticides, and cash that come along guidance are offered to the farmers to abandon the illegal cultivation.

Effects of Plantation

The plantation afforestation in Bangladesh has severe consequences. The kinds of plantations in Bangladesh are basically 'simple' monoculture that wipes out the native species in the first place. Opposed to 'simple plantation forestry' "complex plantation forestry offers the prospect of more effective conservation of forest genetic resources, particularly through the *circa situ* conservation" (Sensu Cooper at al. 1992 in Woodwell 2001:108). Although many forest officials express their preference for mixed or complex plantation, in reality we see little of it. As a consequence, our public forestland has actually dried up. The Chittagong Hill Tracts with their sensitive landscapes has witnessed unprecedented damages due to the plantations.

The negative and multiplier effects of man-made forest or plantations are almost similar around the globe, best illustrated in the words of Patricia Marchak, a Canadian scholar: "Plantations are monoculture, and the lack of biodiversity is of concern. They typically have sparse canopies and so do not protect the land; they cause air temperatures to rise, and they deplete, rather than increase, the water-table. They are generally exotic to regions. While the initial planting may be free of natural pests and diseases, that situation will not last, and plantation regions may not be in the position to combat scourges yet to arrive" (Marchak 1997:10).

The headman of Alekhyong *mouza* in Naikhongchhari Cho Hla Mong is very upset as the invasion of rubber and outsiders have devastated the forest and livelihood means of the Marmas and the Chaks in his *mouza*.   © Philip Gain

As we see the plantations on the public forestland in the CHT and elsewhere, the contention of Patricia Marchak becomes clear. It is obvious that one can plant trees but it is plainly impossible to create a forest. Hundreds of species of trees and bushes and a large number of other vegetable species grow at all stages and on the forest floor. The knowledge of the forest dwelling communities, their traditions, culture, history, education—are all part of a forest. All these are affected and damaged in many instances.

The displacement of human communities is another consequence of a plantation economy. In the CHT pulpwood and other industrial plantations (including rubber and tobacco) have displaced human communities (*jumias* in particular) from their roots who have lived in the forests for centuries. Once an area is brought under plantation the customary rights of the local communities is nullified in most part. What are also eroded in this process are the lifestyle, culture, knowledge, history, and traditional values of the forest-dependent communities. Total devastation takes places during the harvest time of plantations. This is a clear case of ecocide best illustrated in the CHT [and *sal* forest areas].

Ecological damage caused by rubber plantation is unprecedented. Although a rubber plantation looks green, it is a desert to other plant species, birds, and wildlife. In a rubber plantation one feels like being in a green desert. It brings some cash for the government and private entrepreneurs, but misery and trouble for the local communities. It destroys the grazing land for the villages around. It offers some jobs to the locals as tappers and day laborers, but not without serious consequences. In Bandarban where rubber has invaded the *jum* land, the local indigenous communities have lost access to their traditional land.

In the CHT plantations have provided ample grounds to the land grabbers. In most cases it is the rich, influential or outsiders who have encroached upon the forestlands in collusion with the government agencies and political forces.

References

A Brief on Bangladesh Forest Industries Development Corporation (BFIDC). 2005.

Ahmed, Q.K.et al (eds). Resources, Environment and Development in Bangladesh. p. 54. (Note: Forests in this section on Bangladesh include natural forests and plantations).

Alam, M.K and S.K. Khisa. 1998. Preliminary Studies in Ethnobotany of Chittagong Hill Tracts, Bangladesh and Its Linkages with Bio-diversity. A discussion paper submitted to ICIMOD, Kathmandu, Nepal.

Amar Desh, 31 May 2008. *Deshe Tamaksrishto Aat Roge Bochhore Mara Jay Mote Mriter Shola Bhag* (Every year 16% of total death occurrence is caused by eight tobacco-related diseases in Bangladesh).

Anti-Slavery Society (Indigenous Peoples and Development Series: 2). 1984. The Chittagong Hill Tracts: Militarization, Oppression and the Hill Tribes. (The name of the organization is not slightly changed as Anti-Slavery International).

Arens, Jenneke. Foreign Aid and Militarization in the Chittagong Hill Tracts (1997) in "Living on the Edge: Essays on the Chittagong Hill Tracts" edited by Subir Bhaumik, Meghna Guhathakurta and Sabyasachi Basu Ray Chaudhury (ed.). South Asia Forum for Human Rights, Calcutta Research Group.

Arnab, Shanjid. *Tamak Chas: Humkite Jonoshthyo o Khaddo Nirapatta* (Tobacco Cultivation: Public Health and Food Security Threatened. *Saptahi 2000,* 6 May 2011.

Asian Development Bank (ADB) and People's Republic of Bangladesh. Chittagong Hill Tracts Development Project, Main Report, 1978.

Asian Development Bank (ADB) and People's Republic of Bangladesh. Chittagong Hill Tracts Development Project, Annexes 1-8, Background to the Project, 1978.

Asian Development Bank (ADB) and People's Republic of Bangladesh. Chittagong Hill Tracts Development Project, Annex-4, pp.10, 26. 1978.

Asian NGO Coalition for Agrarian Reform for Agrarian Reform and Rural Development (ANGOC), Philippines and International Land Coalition (ILC), Rome Italy. 2006. Enhancing Access of the Poor to Land and Common Property Resources.

Bangladesh Bureau of Statistics (BBS). Statistical Year Book of Bangladesh, 1997.

BBS, 1996. Women and Men in Bangladesh: Facts and Figures.

BBS. Bangladesh Population Census 1991. Analytical Report, National Series.

Bhaumik, Meghna Guhathakurta and Sabyasachi Basu Ray (ed.) South Asia Forum for Human Rights, Calcutta Research Group.

Brammer, H, 1986: Reconnaissance Soil and Land Use Survey: Chittagong Hill Tracts-1964-1965, Soil Resources Development Institute, Dhaka

Brauns, Claus-Dieter and Löffler, Lorenz. 1990. Mru, Hill People on the Border of Bangladesh (Revised Edition in English).

Chakma, Hari Kishore. *Tin Parbotto Jelar Pahari Bhumite Ganja, Poppy O Afim Chash, Koti Taka Aye Korchhe Osadhu Beboshai.* (Cultivation of Ganja, Poppy and Opium in Three Districts of Chittagong Hill Tracts, Corrupt Businessman Earns Millions) *Prothom Alo*, February 9, 2003.

Forest Department. 2001. *Ekush Shatoker Bono Bibhag* (Forest Department of 21st Century).

Forestal Forestry and Engineering International Limited. May 1966. The Chittagong Hill Tracts Soil and Land Use Survey 1964-1966. Volume 5, Forestry—Land Use.

Gain, Philip (Ed). 1998. Bangladesh: Land Forest and Forest People. Society for Environment and Human Development (SEHD)

Gain, Philip (Ed). 2000. The Chittagong Hill Tracts: Life and Nature at Risk. SEHD.

Gain, Philip and Moral, Shishir. Land Right, Land Use and Ethnic Minorities of Bangladesh. Land, December 1995, Vol. No.2.

Gain, Philip. 2006. Stolen Forests. Society for Environment and Human Development (SEHD), Dhaka.

Halim, Sadeka; Roy, Raja Devasish; Chakma, Susmita; and Tanchangya, Sudatta Bikash. 2007. Bangladesh: The Interface of Customary and State Laws in the Chittagong Hill Tracts in BRIDGING THE GAP: Policies and Practices on Indigenous Peoples' Natural Resource Management in Asia. UNDP-RIPP•AIPP Foundation, Chiang Mai, Thailand.

Hasnat, Abul. *Tamakchullite Terolakh Gachh* (1.3 million trees in Tondur). *Prothom Alo,* 5 May 2011.

Hossain, Dr. A.T.M. Emdad. *The Role of Private Sector Rubber Plantation in Economic Development of the Country* (translated from Bangla) presented at the annual meeting of Bangladesh rubber plantation owners' association on 16 February 2013 at Bangladesh Forest Research Institute, Chittagong.

http://womenshealth.about.com/cs/azhealthtopics/a/smokingeffects.htm. Accessed on 27 February 2012.

Hutchinson, R.H. Sneyd. *Eastern Bengal and Asam District Gazetteers: Chittagong Hill Tracts* (Aalahabad: Pioneer Press, 1909).

Ishaq, Muhammed. 1975. *Bangladesh District Gazetteers: Chittagong Hill Tracts, 1971*. Ministry of Cabinet Affairs, Government of Bangladesh.

Kanowski, Peter (in Woodwell, George M. *Forest in A Full World*, Yale University, USA 2000).

Khan, Md Habibur Rahman. Bangladesh Forest Industries Development Corporation (BFIDC). Bangladepdia, National Encyclopedia of Bangladesh, March 2003.

Löffler, L.G. 1991. Ecology and Human Rights: Report on a short visit to Bandarban and Rangamati (Chittagong Hill Tracts, Bangladesh).

Marchak, Patricia M. 1995 (reprinted 1997). Logging the Globe. McGill-Queen's University Press.

Mohsin, Amena. 2000. State Hegemony [in the CHT] in The Chittagong Hill Tracts: Life and Nature at Risk edited by Philip Gain.

Partap, T, 1998: Sloping Land Agriculture and Resource Management in the Semi-Arid and Humid Asia: Perspectives and Issues. In APO (ed): Perspectives on Sustainable Farming Systems in Upland Areas, Asian Productivity Organization, Tokyo, Japan.

Prothom Alo 31 May 2011. Lokshane Komechhe Lover Tamak Chash (Tobacco cultivation of greed reduces due to loss).

Prothom Alo, February 27, 2005. Thanchite Bon Ujar Kore Bistirno Elakay Poppy Chash (Massive Forest Destruction by Poppy Cultivation in Thanchi).

Roy, Raja Devasish and Gain, Philip. 1999. Indigenous peoples and forests in Bangladesh in Forests and Indigenous Peoples of Asia. Minority Rights Group.

Roy, Raja Devasish. 2002. Land and Forest Rights in the Chittagong Hill Tracts, Bangladesh. International Centre for Integrated Mountain Development (ICIMOD), Kathmandu, Nepal.

The Daily Star February 25, 2011. Poppy Fields Destroyed in Bandarban.

The Daily Star, 31 October 2002. Tobacco Replaces Crops, Vegetables in CHT Hills.

The Daily Star, February 25, 2011. Poppy Fields Destroyed in Bandarban.

The Report of the Chittagong Hill Tracts Commission. 1991. 'Life Is Not Ours': Land and Human Rights in the Chittagong Hill Tracts, Bangladesh.

Timm, Father R.W. 1991. Adivasis of Bangladesh. Minority Rights Group International, London.

Van Schendel, Willem; Mey, wolfgang; and Dewan, Aditya Kumar. 2000. The Chittagong Hill Tracts: Living in a Borderland. White lotus, Bangkok.

Whitemore, T.C. 1998. An Introduction to Tropical Rain Forests (Second Edition).

Woodwell, George M. et al. Forests in a Full World. Yale University 2001.

World Rainforest Movement (WRM). Ten Replies to Ten Lies.

Note: Annya Kor, an intern with SEHD, assisted the author in collation of information on tobacco and poppy from secondary sources.

The Disastrous Kaptai Dam

Raja Tridiv Roy

Spillway at Kaptai. Courtesy: Bientjes, 1962 in Schendel,
Mey & Dewan, 2001, The Unversity Press Limited.

In 1782 Warren Hastings was appointed Governor of Bengal and the following year he became the first Governor General, under Lord North's Regulating Act, with a Council of four members. In 1784 he resigned. He went back to England, was impeached in Parliament and seven years later was acquitted. For a year McPherson held the fort. Lord Cornwallis took over in 1786 and ruled India until 1793.

For a number of years the Chakmas waged war against the English and bested them in a number of battles. But when his people suffered severe privations due to and economic blockade Jan Bux opted for a peaceful resolution. Since the Chakma King Jan Bux Khan had signed a peace treaty with Lord Cornwallis at Fort William, Calcutta, in 1787, the biggest single calamity for the Chakmas up to 1960, has been the dam - the hydroelectric project at Kaptai. Soon after the creation of Pakistan in 1947, the East Pakistan government revived the long-abandoned British scheme for a hydroelectric project at Barkal, 28 miles above Rangamati. There were natural rapids here below a deep gorge but the site was rejected ostensibly due to its proximity to the Indian border. If the dam had been built here some Indian territory, mainly barren hill slopes, would no doubt have been submerged. Navigation in the upper reaches of the Karnafuli River in India itself, and its tributaries, like the Harina and the Thega on the border, and tributaries upstream however, like Harani-Marani would have greatly improved, and possibly India would have agreed to the loss of a few square miles of infertile hill terrain. At best she would have demanded some compensation. However, Government opted not to negotiate with India.

Foreign engineers and experts then selected another site, a steep rock face at Chilardhak, 20 miles downstream from Barkal. Inexplicably, after some initial work, this site also was abandoned. Instead, the East Pakistan government selected a site downstream well below Rangamati. The hill men felt the main reason for constructing the dam at Kaptai was politically motivated. It was a well deliberated decision to disperse the bulk of the Ckakmas, particularly the well to do and the influential that inhabited the river valleys of the Karnafuli and its main tributaries, the Chengi, the Subalong and the Kassalong. The motive, they felt, was to disperse and weaken the most advanced, the Chakmas, politically and economically. It was called a multi purpose dam, for it was supposed to provide not only electricity but flood control in the plains of Chittagong and irrigation facilities as well. As it turned out, since the dam was built there have been floods in the very region it was supposed to have saved, with almost unfailing regularity. As for irrigation, by its very coming into existence it submerged 54,000 acres of scarce plough lands. There was hardly anything left to irrigate except outcrops of rock or shale or stretches of wild bamboo and, here and there, steep sandy hill slopes, which needless to say required no irrigation, and unwise disturbance of the topsoil by buffeting waves and the fluctuating lake level only accelerated erosion.

Factually, there were two benefits besides the generation of electricity that however, only provided a very small proportion of the province's requirement — improvement in navigation, though the rate of siltation has been more rapid in the upper reaches of the dam than expected, and fishery. Needless to say, despite of repeated and grandiose government plans and promises, not a single indigenous village has been electrified-though electricity found its way into towns, as well as villages in the plains districts. Nor has any part or percentage of the income from the sale of power been provided to the Hill Tracts. It is a repetition of the case of the Paper Mills at Chandraghona, where the hill people have neither been given employment nor is any portion of the income spent on the Hill Tracts, though all the raw material is supplied by the Hill Tracts. This is very different from the natural gas from Sui, in the Bugti Area of

Balochistan, where there is an agreement with the tribe and substantial sums by the gas extracting corporations and companies are paid every month to the tribes to be spent as they please.

Original estimates indicated about 50,000 persons would have to be shifted and about 200 square miles of cultivable as well as hill lands would be inundated. In the mid 1950's, the government set up a committee under my chairmanship, which included one or two government officers, to give our views. We submitted a comprehensive report. We estimated on the basis of data provided to us that about 87,000 people would be displaced due to inundation, and about 30,000 others would also have to shift if the 87,000 shifted. If, for instance, 80% of a village had to leave, the remaining 20% could not exist in isolation, marooned in small isolated and dispersed islands, with the existing facilities all gone, including their only source of livelihood-the cultivable plough lands.

We made number of recommendations with regard to alternative lands for rehabilitation, mode and phases of shifting, mechanics of assessment and payment of compensation, and the economic rehabilitation of the displaced. We also proposed measures to ensure that integrity of the indigenous people was maintained. Many committees at various levels were subsequently set up and I had serious differences of opinion, which I voiced at these committee meetings. The government, either by design or by default, dealt with the problems most cavalierly, siphoned off the funds (over $ 51 million received from abroad specifically for this purpose), allocating a paltry sum of $ 3 million for rehabilitation and compensation. They entrusted the operations to hidebound or corrupt officials. There were a few notable exceptions like Syed Afzal Agha, M.A Kareem Iqbal, and O.M.Qarni-C.S.P officers from West Pakistan. The man in overall charge was an old ICS (Indian Civil Service) officer called Hatch-Barnwell, Member Board of Revenue. The nature of the task, however, was too gigantic for one or two men, especially when hamstrung with in unsympathetic government.

Syed Afzal Agha was transferred out due to his sympathy for the hill people, and O.M. Qarni, the Rehabilitation Officer cum Additional Deputy Commissioner's views were likewise ignored. The brunt of the thankless task fell on Kareem Iqbal, who with great Limitations of funds did his best. He was an extremely hardworking and efficient officer trying to build a pyramid with pebbles almost single-handed. The maximum compensation for land, for instance, was arbitrarily fixed at Rs. 900 per acre and the minimum at Rs. 300. This was an outrageous figure because it was hardly the price of one year's crop. A mango tree was assessed at Rs. 10 ($ 2 at that time) although one season's crop fetched at least ten times that amount, and even if the tree were to be sold as firewood it would have fetched more.

Shifting expense was fixed at so ridiculous a figure that I refused to accept any money for myself and my staff and dependants. I told the government that they paid more money to their *peons* and *chaprassis* (low-paid office orderlies) when they were transferred from one station to another. For lands, houses and tress I cited rules and practices of government but as this was an Excluded Area those rules did not apply. Yet when it suited them the "Excluded Area" status did not prevent them from introducing adult franchise in order to bring the Tracts within the ambit of the provincial legislature or in introducing the harsh and discriminatory Agricultural Income Tax Act. But what the provincial government did apply was grotesque and monstrously iniquitous.

Along with other leaders, we made innumerable representations and took delegations to government functionaries, including the governor of the province. We heard soothing words and many promises of redress but in substance very little was gained. If we had guns (and hill men were given even shotgun licenses with great reluctance), we would have fought even unto extermination. That was the feeling of the vast majority. We had no guns so we seethed in silence, in humiliation and anger.

To cap it all, even from the meagre pittance given by government for "rehabilitation and compensation", a good portion was misused (which was expected), misappropriated (naturally), and even in making payment to the displaced persons some of the disbursing officers retained a certain percentage. If anyone demurred, he was fobbed off and told to come on another day and so on and on, *ad infinitum*. One Chakma received some money-minus the illegal extortion and in front of the officer tore the currency notes into bits and threw it at the officer's face. The mealy-mounthed hypocrites said it was to ensure that the simple people didn't "squander their money on luxury items" that they had decided on installment payment, overriding our protests. On the next installment also the fixed percentage was retained from the man's due compensation. He complained against the official but there was no proof, because you sign first and receive after, inside a room where there is only officialdom.

Government admitted that more than 1,00,000 persons were displaced. The area inundated is estimated to vary from 400 to 600 square miles, in place of 297 square miles as had been officially given out. The level in the lake, in the course of a year, fluctuates more than thirty feet and some lands are out of water for a part of the year. Government inefficiency reached its pinnacle in having to re-rehabilitate, that is to say, twice displacing several thousand persons who were flooded out of their new government allotted lands in the Sharbuatali and Khedarmara areas of southern Kassalong Rehabilitation Area. The proffered excuse-faulty contour marks on old survey maps.

From amongst those shamans who knew the ancient magical formulas, one or two are said to have used their art on the dam site at Kaptai and caused epidemics like cholera, twice. They also claimed they succeeded, through the intervention of malleable sprits, in cracking a portion of the coffer dam when it was halfway across the Karnafuli, thereby putting back the work schedule by another year. This did happen though the cause may have been different. The hapless can always find solace in imagination. Against cruelty and tyranny, whom else could the Chakmas count on but spirits?

The old Rajbari stands about thirty feet under water. The Karnafuli Dam at Kaptai inundated the catchment area partially in 1960 and fully by 1962. The dam affected directly and indirectly, almost entirely the Chakma territory that comprises half the Hill Tracts, and a small area near Mahalchari of the Mong Raja's Circle. Over 100,000 people, preponderantly Chakma, were displaced from their lands and their homes, and over 54,000 acres of scarce arable land (40% of total cultivable land in the Hill Tracts) was lost to the waters of the lake. The rehabilitation of some of the affected plough cultivators, and nominal compensation by government was, to put it mildly, hopelessly inadequate. It is a verified fact that bulk of the World Bank and other funds received from international sources for the purpose was diverted elsewhere by the government, as confirmed by Prof. Amena Mohsin in her writings and monographs, and others.

The Jummas in the cathchment are that comprised hundreds of square miles, that is, people who lived by jumming, were totally excluded from the Rehabilitation Programme and received neither any monetary compensation nor alternate lands for jumming or fruit growing on the hill slopes. Alternative plough land was of course out of the question as there was not enough available even for a small fragment of those who had lost their cultivated plough lands. Homeless and without resources, felling abandoned, deprived and driven out, over 40,000 displaced persons, almost all Chakma, left for India and took shelter in the remote wilds of Arunachal Pradesh bordering China, then called the NEFA (North

Eastern Frontier Agency). Despite a judicial verdict in their favour, and after a lapse of forty odd years, they have yet to be recognized as Indian citizens. Pradhir Talukder has made a CD video cassette in 2002, with live interviews of some of the Chakmas and others in Arunachal. A retired senior government officer Utankamani Chakmas, who was Secretary to the NEFA government at the time of the Chakma Exodus in 1964 has made a few cogent observations that:

1. Most of the areas where the Chakmas were settled are no man's land.

2. Hundreds by the day perished on the journey to NEFA (now Arunachal) because of malnutrition mainly.

3. He was asked by the Governor and Chief Minister to find places for settling the Chakmas. The areas selected were uninhabited, still the consent of the local tribes living in the vicinity were taken.

Other emerging facts:

4. Kunkugam Singpho also in his interview confirmed that he and his people were happy to welcome the Chakmas, who were also Buddhists like them, and who were in trouble.

5. On 10 February 2002 a whole line of Chakma women and girls were seen participating in the funeral rituals of a senior Buddhist monk and on the 19th many Chakma people were assembled at Diyun Bazar. The Chakma women were selling vegetables, quite some distance away and one middle-aged woman was ploughing the field with a pair of oxen.

6. We also learnt that there is only one doctor for 10,000 people in the three districts of Arunachal, which is about 30,000 sq. miles in area.

7. A Chakma teacher of a primary school mentions that the community pays him and the other teachers. The Government pays the salary of only one teacher.

8. Utankamani Chakma also says that one or two of the smaller non-Chakma ethnic groups are on the verge of extinction and that some had recently come from Burma. There were about a 100 groups of indigenous peoples in Arunachal.

9. At present there were said to be only about 26,000 Chakmas in three districts of Arunachal.

Source: *The Departed Melody* (memoirs) by Raja Tridiv Roy Published by PPA Publications, Islamabad, 2003.

Serfdom in the Colonial Reserves

Devasish Roy

Introduction

Indigenous communities have been struggling for long to protect their traditional resource rights in the "reserved forests" of the Chittagong Hill Tracts (CHT). These state-owned reserves cover about a quarter of the CHT, and are administered by the national *Department of Forest*. They contain some of the most rugged and remote parts of the country, which have been the home of various swidden or "jum"-cultivating indigenous communities for centuries. A large part of the reserves shares boundaries with India and Burma (Myanmar). Recently, these reserves have also been the shelter of internally displaced indigenous people who left their homes and farmlands during the situation of internal conflict in the CHT in the 1980s. The reserves contain the largest part of the country's biodiverse natural forests, apart from the Sundarbans, although their extent is constantly decreasing.

The Colonization of the Forest Commons

The forest reserves of the CHT are still administered by the Forest Department in more or less the manner they were when these areas were first designated as "reserved" forests in the 1880s. The then colonial government had unilaterally acquired these sparsely populated and densely forested lands to meet the industrial needs of the empire, totally disregarding the indigenous peoples' rights over the concerned lands (Roy & Gain 1999, Roy & Halim 2001a). No indigenous institution or local authority has any say over the management of these lands except as regards the maintenance of law and order, the administration of criminal justice, and the resolution of civil and customary law disputes. The major difference between the British colonial period (1860-1947) and the post-colonial period is that the extent of actual forest cover during the former period was far greater, as the population pressure was low, and policing was stricter. The population of these areas has since risen manifold, despite strong governmental discouragement, and on occasions, heavy-handed action (Roy & Gain 1999). In fact, in accordance with the Forest Act of 1927, adopted during the British period, it is a penal offence to cultivate lands, use forest produce, or to even enter into a reserved forest without the consent of the Forest Department. No ownership or tenurial rights of any inhabitants of the areas are acknowledged by the Forest Department.

Besides the non-acknowledgement of their land and resource rights, the inhabitants of the reserved forests are also denied adequate access to publicly-funded schools and health services, along with other basic government extension services (Ibid: 21-23). They therefore remain as little more than modern-day serfs. This is largely because the Forest Department essentially remains as a quasi-police body that is oriented around administrative, rather than extension, services, drawing upon a system that dates back to the 19th century, during British colonial rule.

Servitude in the State Plantations

Ironically enough, at times, the Forest Department had even encouraged limited migration of indigenous people into these areas, intending thereby to secure their cheap labour for the creation of mono plantations of teak or other alien and local species of trees of high industrial or commercial value. Such migration took place especially in the post-colonial period after 1947. In addition, some indigenous communities used to live within these areas well before their categorization as government forests. However, the Forest Departments of both the Pakistan period (1947-1971) and the Bangladesh period (1971+), have always been very careful to avoid any formal recognition of these communities' rights over their homesteads and surrounding agricultural lands, let alone acknowledge their rights over the new plantations that their hard labour had helped create! This has been so despite the presence of some provisions in the otherwise *statist* Forest Act of 1927 that provides a scope for limited recognition of community title over reserved forest settlements that are regarded as "forest villages" (Roy & Halim 2001a).[1] To make matters worse, large numbers of people from these settlements have also continually suffered from oppressive and false criminal charges and other discriminatory and occasionally violent acts at the hands of police and armed Forest guards. Such concerted acts were ostensibly aimed at the protection of rising theft of timber from these protected areas, but actually were no more than attempts on the part of Forest Department employees to escape personal and departmental responsibility for the loss of government property, in which they themselves were usually implicated (Ibid). Recently, the paramilitary Border Guard Bangladesh (BGB), formerly *Bangladesh Rifles* (BDR), is also reported to have facilitated the eviction of *jumia* farmers from *jum* plots within the northern Kassalong reserve, home to the largest concentration of internally displaced indigenous people in the region.

The Reserved Forest Communities Assert their Rights

In *some* reserved forest areas, the indigenous people were able to exercise their voting rights and participate in local council elections for the first time only as recently as 2003, almost half a century after their relatives in the mauza-circle (non-reserved forest) areas of the CHT had secured their franchise rights! And this seemed to please neither the local Forest Department officials nor the local army officials posted near the new Union Council member's village. The question of the forest-dwellers' rights and oppressive situations have been raised in various fora by Taungya (a CHT-based development NGO) and its networking allies, including the *Movement for the Protection of Forest and Land Rights in the CHT* (Roy & Halim 2001a).

It has been argued that the traditional communal and custom-based land rights of the indigenous peoples cannot be extinguished by the 1927 Forest Law, since custom is a part of the Bangladeshi legal system as acknowledged by both the national constitution of 1972 and the CHT Regulation of 1900 (that screens and regulates the application of all laws to the CHT region, including the Forest Act of 1972 itself) (Roy 2004b). Moreover, the provisions of the 1927 Act must be read in conformity with the fundamental rights of citizens as declared in the national constitution since 1972. That, however, is the formal constitutional position. The actual enjoyment of those rights is quite another matter, and may require administrative, legal, and policy changes. Therefore, there are probably few practical alternatives to strengthening the organizational capacities and networking and lobbying alliances of these communities, to bring about the required changes.

In the long run, it is unlikely that substantial progress can be achieved by Reserved Forest communities regarding the strengthening of their land and resource rights merely through lobbying, and related *awareness-raising* and *organizational strengthening* to strengthen the communities lobbying work. This is because most of these communities are socially and economically extremely marginalized and disadvantaged, due, among other reasons, to their settlements' "remote" locations. Therefore, concrete measures are necessary to help these communities access public health services, have improved road communications, market their produce, raise elusive capital through easy-term farm and business loans, and prevent oppressive money-lending practices. They are unlikely to go far in securing their land and political rights if they still have to spend a great deal of time and attention to securing their basic livelihood needs. Therefore, accessing the aforesaid welfare services may be as important as to help make the communities aware about their basic legal rights (Sutter 2000). Attention to such welfare needs is also important due to the long periods of state-neglect and discriminatory development policies of successive governments since the colonial period.

Until recently, the Forest Department appointed "headmen" (not to be confused with the more influential and partly traditional *mauza headmen* in the mauza-circle areas) appeared to hold the most influential leadership positions in the reserved forest areas. The only exception was where the elected *union council* system was extended. There is little doubt that the elected community leaders have been able to voice the concerns about their people much more forcefully than the Forest Department-appointed "headmen". This is largely because these forest headmen have to depend upon the Forest Department to retain their offices. Quite understandably, even these "headmen" cannot call their own homesteads their own land, let alone other forest-dwellers.

In the case of one reserved forest community in the mid-southern Reinghkhyong Reserve near the village of Farua, petitions were sent to the government to "de-reserve" a small area solely to enable them to get state subsidies for their school. Such subsidies could not be availed of, in accordance with the concerned Education Department rules, unless the school land was formally registered as the property of the school managing committee. Although the laws clearly allowed such dispensations, the request was denied by the Ministry of Environment and Forest.[2] Consequently, these areas have to do with a few make-shift schools managed from the community's meagre resources, in addition to a few schools hitherto supported by NGOs and now subsidized through a UN-funded project. Another reserved forest community, within the northern Kassalong Reserve, is also known to have recently applied for de-reservation.[3] However, it is unlikely that the Forest Department will agree to such de-reservation. Except to accommodate government-sponsored Bengali transmigrants in the 1980s, the Forest Department has not recently agreed to de-reserve any portion of the reserved forests. In fact, despite expert advice in the mid-1970s to de-reserve small reserves that were difficult to manage (Webb & Roberts 1976), the Forest Department has instead chosen to *increase* its area of reserved forests through a process that was initiated in the 1970s and 80s, threatening indigenous and other local communities with wholesale eviction. Thus, ever since the colonial period, a sense of territoriality and expansionism seems to have guided the department, whose legacies are still to be discarded.

Limited activities of the social organization, Taungya and the Movement for the Protection of Forest and Land Rights in the CHT have managed to help rally these forest communities in a limited manner to raise their voices. The election of an indigenous *union council member* in the remote and picturesque *Reinkhyongkine Lake* area in 2003 for the very first time shows that these communities are more than willing to engage governmental and developmental agencies regarding their rights and welfare. However, some external support to help them overcome basic livelihood needs is perhaps indispensable, given the communities' socio-economic marginality and relatively "remote" location. Of course, any such support needs to ensure sustainability and avoid undue external dependency that is harmful to the communities' long-term welfare and development goals. The *Forest and Land Rights Movement* and a number of indigenous leaders from different parts of the CHT have attempted to support the organizational strengthening work of these communities, but there does not as yet appear to be any detailed plans or programmes to achieve long-term development goals and basic rights.

Having been deprived for long of their basic civil, political, and economic rights, some inhabitants of the *Reinkhyongkine Lake* area are considering to demand the de-reservation of the entire area, which contains scores of settlements, some "degraded" forestlands, relatively fertile swidden commons, and upland grazing commons that are used for locally-innovated forms of herding of wild and semi-wild bison (*bos frontalis*), sometimes cross-bred with local cattle. However, for the same reasons as in the case of the *Farua* and *Kassalong* communities as discussed above, it is unlikely that such demands will be taken seriously by the government. Sadly enough, for reasons not known, the problems of the reserved forest areas were not directly addressed by the 1997 "peace" Accord on the CHT. The inhabitants of these areas cannot, therefore, invoke the Accord to lobby for their land rights other than in a general way. They are in many respects far more deprived than the inhabitants of the mauza-circle areas. It is ironic that they should have to suffer from policy neglect both from the national government and from their own regional political groups. Their life of serfdom is still far from over, even in the beginning of the 21st century, more than half a century after slavery was formally abolished by international human rights law.

Revisiting the People's Strategies to Protect their Rights

Despite numerous instances of oppressive state action, including arrest, prosecution, etc., the communities have, on the whole, been able to retain their possession, and at least partial control, of their homesteads and farmlands. At the same time, they have also managed to substantively evade expulsion from the areas concerned. In fact, the number of inhabitants of the reserved forests has actually increased, rather than decreased, since the 19th century, and accelerated after the displacement caused by the Kaptai Dam in 1960, the resettlement of Bengali settlers in the 1980s, and anti-insurgency military "operations" during the same period (IWGIA 2004: 295).

The basic rights issues of the inhabitants of the reserved forests have received very little attention at the national level. There is, therefore, considerable scope to enhance work on mobilization, advocacy, and networking. Judicial and human rights-related remedies have been the least used invoked so far, both within the country and in international processes. A major factor behind this is financial and logistical constraints, in addition to technical and informational difficulties.

Seeking Judicial Redress under National and International Law

To what extent the efforts of the forest-dependent communities in the Chittagong Hill Tracts will succeed or fail in protecting their traditional resource rights will of course depend upon numerous influences and processes, both at the national and supra-national levels. It is important, in particular, for the indigenous people and their allies to minutely study the national Forest Policy and the Forestry Master Plan, and to continue to lobby for amendments as necessary. Despite its dysfunctionalities and other difficulties, the strengthened CHT self-government system no doubt has a larger say than before in helping steer relevant government policy, at least by acting as a watchdog in vetting nationally proposed schemes, projects, and programmes. Thus it is extremely important to bring about a close and positive working relationship between organizations working with the forest communities and regional political groups and civil society organizations. Legacies of previous intra-indigenous political differences have on occasions prevented such a relationship, but unless these are overcome, the defence of the traditional resource rights of the indigenous peoples will remain less than truly united. Moreover, as a matter of tactic, it is well to remember that the situation varies between the reserved forest areas and other areas. The former still continue to remain as colonial enclaves, since the 1997 CHT Accord did not directly address the issue of the indigenous peoples' rights over these stolen commons. Therefore, the writ of the customary resource rights system will most likely continue to be rejected by Forest Department officials in the case of the reserved forests, and will therefore require greater attention for concerned activists.

Among other important factors that traditional resource rights defenders in the CHT need to account for are supra-national influences upon the trade and aid-dependent political economy of Bangladesh, including the phenomena of privatization and marketization, and the influence of multilateral development institutions as the World Bank, the IMF, the IFC and the Asian Development Bank, who continue to play a vital role in fiscal and developmental policy-making in Bangladesh. The relevant policies of the World Bank and the Asian Development Bank, for example, on the matter of forestry and on the subject of indigenous peoples, do not favour the strengthening of traditional resource rights. The World Bank's equivocal stand on traditional resource rights of indigenous peoples was apparent in its support to the controversial Tropical Forestry Action Plan and in its new Indigenous Peoples policy (OP 4.10). As for the Asian Development Bank, its own policy on forestry, its role in helping formulate the Forestry Master Plan of Bangladesh and a number of controversial amendments to Bangladeshi forest laws in 2000, in addition to its loans on Forestry Sector Project, have clearly demonstrated its real and potential negative impacts on the protection of forest commons and the strengthening of community participation and traditional resource rights (Roy 2002a; Roy & Halim 2001a). The contents of and the nature of the implementation of the relevant policies of the aforesaid institutions are therefore key factors in influencing future trends with regard to the situation of traditional resource rights of indigenous peoples, such as those living in the Chittagong Hill Tracts, and in comparable areas elsewhere in the world.

The situation of traditional resource rights of indigenous peoples in the Chittagong Hill Tracts and in many other parts of the world will also depend, to a great extent, on other existing and emerging international standards and processes including those on trade (WTO), and on intellectual property rights (WIPO). Other crucial factors that will influence the situation of traditional resource rights of the indigenous peoples of the CHT are the nature of the implementation or non-implementation of relevant international treaties. These include the ILO Convention on Indigenous and Tribal Populations (No 107 of 1957), the International Convention on the Elimination of Racial Discrimination (CERD), the Convention on Biological Diversity, and the "human rights treaties" (the International Covenant on Civil and Political Rights, and the International Covenant on Economic, Social and Cultural Rights), all of which have been ratified by Bangladesh. Among these, the ILO Convention No. 107 and the Biodiversity Convention are the most directly related to traditional resource rights.

Another crucially important matter for indigenous peoples worldwide, is the proposed adoption of a declaration on indigenous peoples' rights by the United Nations. An initial draft was adopted by the UN Sub-Commission on Human Rights in 1994. The draft contains many crucial provisions on indigenous peoples' collective and individual rights, including traditional rights over forests and other commons. A small number of rich and powerful industrialized countries, ironically enough, also being those who carry the legacies of a history of colonization and oppression of indigenous peoples, continue to stand in the way of such adoption, with strong reservations towards collective rights, amongst others.[4] This does not augur well for indigenous peoples and their communities' traditional resource rights, and calls for concerted action worldwide, to counter such narrow, parochial and undemocratic efforts. This will not be easy.

Notes

[1] Section 28 of the Forest Act of 1927 allows the state to recognize the community title of "forest villagers" in selected parts of reserved forests

settled by local people, a provision not known to have been invoked in the CHT, and only very selectively invoked in the northeastern Sylhet administrative division within Bangladesh, in the case of indigenous Khasi communities through the execution of limited period leases, but later discontinued.

[2] There were a series of negotiations in 1999 and 2000 between the local communities in the Farua Union Council area with district-level Forest Department officials, with some facilitation by Taungya, to obtain de-reservation of the school land (which ultimately came to nought). This information was provided to the author by workers of Taungya and community leaders of the area concerned.

[3] Information provided to this author by Thano Chandra Karbari, a community leader, in several interviews from 1999 to 2003.

[4] The aforesaid opinion is based upon the writer's observation and direct participation in the meetings of the UN Working Group on the Draft Declaration for the last few years. He hopes and expects to attend the forthcoming meeting of the *Working Group* to be held in Geneva in November-December 2005.

Bibliography & References

Adnan, Shapan. 2004. *Migration, Alienation and Ethnic Conflict: Causes of Poverty in the Chittagong Hill Tracts of Bangladesh*, Research & Advisory Services, Dhaka.

IWGIA. 2004. *The Indigenous World: 2002*-2003, Copenhagen.

Larma, Jyotirindra Bodhipriyo. 2003. *"The CHT and its Solution"*, paper presented at the Regional Training Program to Enhance the Conflict Prevention and Peace-Building Capacities of Indigenous Peoples' Representatives of the Asia-Pacific" organized by the United Nations Institute for Training and Research (UNITAR) at Chiang Mai, Thailand on 7-12 April 2003.

Lynch, J, Owen, and Kirk Talbott. 1995. *Balancing Acts: Community Based Forest Management National Law in Asia and the Pacific,* World Resource Institute.

Minority Rights Group International. 1999. *Forests and Indigenous Peoples of Asia*, Report No. 98/4, London, 1999.

80   The Chittagong Hill Tracts: Man-Nature Nexus Torn

Mohsin, Amena. 2003. *The Chittagong Hill Tracts, Bangladesh: On the Difficult Road to Peace*, International Peace Academy Occasional Paper Series, Lynne Rienner Publishers, Boulder, London.

Roy, Raja Devasish. 2004a. *Background Study on the Chittagong Hill Tracts Land Situation*, CARE-Bangladesh, Rangamati.

_____. 2004b. "Challenges for Juridical Pluralism and Customary Laws of Indigenous Peoples: The Case of the Chittagong Hill Tracts, Bangladesh" in *Arizona Journal of International and Comparative Law*, Vol. 21, No. 1, published by James E. Rogers College of Law, the University of Arizona, pp. 111-182.

_____. 2003. "The Discordant Accord: Challenges towards the Implementation of the Chittagong Hill Tracts Accord of 1997", in *The Journal of Social Studies*, "100th Issue: Perspectives on Peace: Visions and Realities", Centre for Social Studies, Dhaka April-June, 2003, pp. 4-57.

_____. 2002a. "Perspectives of Indigenous Peoples on the Review of the Asian Development Bank's Forest Policy", paper presented at the *Regional Workshop on Review of the Asian Development Bank's Forestry Policy*, organized by the Asian Development Bank in Manila, Philippines on 14-15 February 2002.

_____. 2002b. *Land and Forest Rights in the Chittagong Hill Tracts*, "Talking Points", 4/02, International Centre for Integrated Mountain Development (ICIMOD), June 2002.

_____, 2000a. "Occupations and Economy in Transition: A Case Study of the Chittagong Hill Tracts", in *Traditional Occupations of Indigenous and Tribal Peoples*, ILO, Geneva, pp. 73-122.

_____. 2000b. "Administration", in Philip Gain (ed.), *The Chittagong Hill Tracts: Life and Nature at Risk*, SEHD, Dhaka, pp.43-57.

_____. 1994. "Land Rights of the Indigenous Peoples of the Chittagong Hill Tracts" *in* Shamsul Huda (ed.), *Land: A Journal of the Practitioners, Development and Research Activists*, Vol. 1, No. 1, pp. 11-25, Dhaka, February 1994.

Roy, Raja Devasish & Sadeka Halim. 2001a. "A Critique to the Forest (Amendment) Act of 2000 and the (draft) Social Forestry Rules of 2000" in Philip Gain (ed.), *The Forest (Amendment) Act, 2000 and the (draft) Social Forestry Rules, 2000: A Critique*, SEHD, Dhaka, pp. 5-45.

_____. 2000b. "Valuing Village Commons in Forestry:

A Case from the Chittagong Hill Tracts", in *Chittagong Hill Tracts: State of the Environment*, edited by Quamrul Islam Chowdhury, Forum of Environmental Journalists of Bangladesh (FEJB), Dhaka, September 2001.

Roy, Raja Devasish & Philip Gain. 1999. *Indigenous Peoples and Forests in Bangladesh* in Minority Rights Group International (ed), "Forests and Indigenous Peoples of Asia", Report No. 98/4, London, 1999.

Sutter, Phil. 2000. "Livelihood Security in the Chittagong Hill Tracts: Findings from a Rural Assessment undertaken by CARE-Bangladesh", Dhaka.

Taungya. 2003. "Second Quarterly Report on Project for the Protection of Village Common Forests in the CHT", November 2003.

Tripura, Prashanta. 2000. "Culture, Identity and Development", in Philip Gain (ed), *Chittagong Hill Tracts: Life and Nature at Risk*, SEHD, Dhaka, pp. 97-105.

Webb, W.E. & R. Roberts, 1976. *Reconnaissance Mission to the Chittagong Hill Tracts, Bangladesh: Report on Forestry Sector, Vol. 2*, Asian Development Bank, May 1976.

Wessendorf, Kathrin (ed). 2001. *Challenging Politics: Indigenous Peoples' Experiences with Political Parties and Elections*, International Work Group for Indigenous Affairs (IWGIA) Document No. 104, Copenhagen.

This article is based upon one section of a paper entitled "The Politics of Safeguarding Traditional Resource Rights in the Chittagong Hill Tracts, Bangladesh: An Indigenous Perspective", that was presented at the Tenth Biennial Conference of the *International Association for the Study of Common Property* (IASCP), held in Oaxaca, Mexico, on 9-13 August 2004 under the panel discussion on "Challenges of Globalizing Commons and Indigenous Peoples of the Eastern Himalayan Region" (12 August 2004).

Major Reserved Forests of the CHT

"The Reserved forests of the CHT are almost ruined," says Goutam Dewan, a frontline leader fighting for protection of land, forest, and people. Needless to say the reserved forests of the CHT have lost their majestic look and features. Reports on four major reserved forests of the CHT depict how the resources of the resereved forests have been plundered.

Reingkhyong Reserved Forest

Partha Shankar Saha

"River Reingkhyong and Kaptai
Matamuhuri and Alikadam
All are mine
All will be."

This song, written in Tanchangya language, narrates the natural beauty of the Chittagong Hill Tracts (CHT) and explains the ties of the Tanchangya people with the rivers, hills, and the land. The Tanchangys are the fifth largest ethnic community in the CHT and the majority live in Bilaichhari upazila in Rangamati Hill District. The Reingkhyong river flows through this upazila. The forest cover on the both sides of the river is Reingkhyong Reserved Forest. Farua, the largest Union Parishad in Bilaichhari occupies two-thirds of the upazila and the entire union is Reingkhyong reserved forest. Eighty percent of 20,000 people in Farua [who live in 50 hamlets] are Tanchangyas.

Reingkhyong Forests: Basic Facts: The Farua Reserved Forest belongs to CHT South Forest Division in the Rangamati circle of 622,187.76 acres.

It has six reserved forests—Reingkhyong, Thega, Suvalong, Sitaphar, Rampahar, and Barkal. The first three are called Reingkhyong Reserved Forest (Dey 2003:3). It is the second largest reserved forest of 188,537.60 acres (The largest one is Kassalong with 340, 552.80 acres).

The British colonial authorities declared a vast area of the hill forest 'reserved' in the eighteenth century. The FD declared 750 square miles of Sitapahar in Maini valley 'reserved' in 1875 and 1345.00 square miles of Matamuhuri, Kassalong and Reingkhoyng valleys between 1880 and 1883 (Dey 2003:3). Earlier in 1920 the FD started raising teak gardens for the first time in Machkumba village in Reingkhoyng forest. The formation of these plantations and the reservation of the natural forest "was one of the first instances in the history of the CHT when the exercise of the hill people's traditional rights in their ancestral homeland came to be regarded as a crime overnight, and for most of them, without any notice" (Roy in *Earth Touch* 1997:35). Thus thousands of acres of natural forest were cleared to give way to plantations of teak and other imported species, "which has not only led to the permanent loss of variety of indigenous plant and animal species of the region, but has resulted in severe soil erosion in these plantations: (Roy in *Earth Touch* 1997:35). This process continued till the 1990s. Later on, the government acknowledged the ill effects of teak on the soil (Bangladesh District Gazetteer, Chittagong Hill Tracts 1971: 6).

The Reingkhyong people still remember the oppression on them during plantations. "We had to do everything—clearing forests, planting saplings, and fetching water from long distances. The FD officials tortured us including hill women," said Tejendralal Tanchangya (76), the UP Chairman of Farua. Most of the people who migrated to Reingkhyong lost their homes and land due to the Kaptai Hydro-electricity Project.

Forest and Forest People in Farua: The FD defines the *jum* as 'traditional' cultivation method resulting devastation of the forests. It blames *jum* for obstructing regeneration of bamboo and accelerating soil erosion. The rivers and the lakes are filled up due to siltation that causes a threat to the Kaptai Hydroelectric Project. The biodiversity is also threatened (The Rangamati Circle at a Glance p. 21).

The Asian Development Bank (ADB) in a report titled *Reconnaissance Mission to the Chittagong Hill Tracts, Bangladesh: Report on Forestry Sector* (by W. E. Webb and R. Roberts, p-13, 1976) estimated that 65% of this forest was destroyed by *jumias* straying into the forest (cited in Roy 1997). The CHT Gazetteer, a government document states, "The forest, particularly the Unclassed State Forest (USF) has been subjected to most ruthless destruction by *jhumming"* (Ishaq 1971:104). However, the tendency of blaming *jumias* for forest destruction is very ancient and colonial. But these documents make no mention of the impacts of the plantation of teak and other exotics on hill ecology. The plantations, in most cases, are established by clearing jungle.

"We did not plant teak clearing the natural forest. We are the silent witnesses of the plundering by FD officials and the influential groups. Is it possible to destroy a big forest by a few *jumias?"* asked Nabarup Tripura, a *jumia* in Tarachhari village of Reinkhyong. The *jumais* alleged the giving of bribes to the FD officials to access the forest for cultivation and it is an 'open secret'. However, the FD officials disagreed.

The DFO of CHT South Forest Division, Subedar Islam, contested Nabarup Tripura: *"Jumias* are encroachers in this forest. There is no doubt *jum* causes immense damage. This is true not only for this forest but for other forests as well. If *jum* cultivation is not contained, the forest cannot be protected." As regards allegation of plunder of trees and association of the FD officials, Islam said: "We do not hesitate to take action if we get specific allegation. There are many other reasons for pillage of the forest resources in addition to *jum* that we cannot spell out. We, the forest officials, have limits."

Philip Gain, a researcher who has intensively observed the forestry sector writes, "... production of hydroelectricity, militarization, Bangali settlement, logging, commercial or industrial plantations and the search for natural resources such as gas and oil are all parts of a cruel process, which is primarily responsible for deforestation and the deteriorating soil conditions in the Chittagong Hill Tracts (Gain 21:2000)."

Forest Case: Baisabi Tanchangya (60) and Laxmimala Tanchangya (72) are two hill women of Ekkuijachhari and Goinchhari villages in Farua union. Earlier in 1981, they were preparing their agricultural land one morning. The army personnel had cut fuel wood in a nearby bush. A few hours later, the FD workers came, seeing the trees cut and accused Baisabi and Laxmimala for cutting the trees and filed a forest case against them. None of these village women had ever been to Rangamati town. They were hiding for a long time after the case was filed. They alleged that the foresters took money from them several times. They do not know whether the case was still pending.

"Forest cases keep the people of Farua in constant anxiety. The Forest Department people steal trees from the forest and accuse the hill people," asserted Na Uchha Marma, a woman councilor of Farua Union Parishad

(UP). "The outsiders steal forest resources but the innocent people are rendered scapegoats. If we had stolen trees we would have been rich. You will not find any rich person among the hill people in Farua."

Health: The single health sub-center of Farua is in Jamunachhari village. There are some medicine shops at Farua Bazar. The health assistants visit twice or thrice a month. The Farua people travel to Bilaichhari Upazila Health Complex or Chandraghona Missionary Hospital to see doctors. Both take three to four hours to reach by river in the rainy season. But during the dry season, it takes five to six hours. Many people do not see a graduate doctor in their lifetime.

The health officer Dr. Masud said they cannot provide proper health care due to the remoteness of the place and the terrible infrastructure. However, the United Nations Development Program (UNDP) along with other NGOs has been working to decrease the death rate of this area due to Diarrhea and Malaria. The Diarrhea and Malaria affected people in Bilaichhari were 2,996 in 2008, which came down to 1,416 in 2009 and 1,309 in 2010. The Diarrhea affected people in upazila were 298 in 2009, 217 in 2010, and 196 till the middle of 2011.

Education: There are three primary schools in Farua but no high school. There is a junior high school without government approval. Goinchhara primary school, one of three schools, has 215 students and three teachers. The headmaster stays busy with administrative work most of the time. So two teachers manage the students.

The NGOs are working in Farua to scale up education. "We have five schools without registration. We had to wait for a long time to get approval," informed Amar Sadhan Chakma, executive director of Taungya, a Rangamati based NGO. Achhya Tanchangya who graduated from Dhaka University is a fortunate man from Ekkujayachhari village in Farua. "I am fortunate enough to have reached this stage. Many students in this area cannot continue their education after primary school as the schools are far away from our village," said Achhya. "Bangla is still the medium of education in primary schools in this area. This is also a reason for the drop-out of children." Additionally, another reason is many students have to go for *jum* cultivation with their parents skipping their studies.

Agriculture: According to Bilaichhari agriculture extension office, the total arable land in Farua UC is 4,153 ha of which *jum* land is 1,677 ha. However *jum* is practiced only on 600 ha of land. In the plains land people depend mostly on *jum*. The *jumias* grow rice, sesame, kaun rice, corn, turmeric, banana, papaya, and many other vegetables. Of late, they are growing mango, litchi, orange, etc.

However, Farua is a food deficit area. "Each year the demand for rice in Farua is 3,070 metric tons while it produces 1,275 metric tons. Due to the poor communications system and lack of marketing opportunity for the products, they have been cultivating tobacco with the assistance of tobacco companies. It is a big threat for local agriculture," said Tarun Bhattacharya, Bilaichhari upazila agriculture officer. "The farmers, without land ownership in the reserved forest run into many difficulties. One major difficulty is they are deprived of agricultural loans."

Farua People Want the Forest De-reserved: According to section 28 of the forest law, the government can allow someone to live in the reserved forest as 'villagers' and can cancel the permission. The Farua people, living in the area for generations, want the forest to be de-reserved to be able to establish land ownership. A part of Kassalong forest was de-reserved in the past for rehabilitation purposes.

"We are outsiders in our own land. We have been living in the forest for a long time; still we don't have any right to this land. We cannot access bank loan having no land ownership. We need permission prior to cultivation. I can plant a tree but I cannot establish ownership to that tree; the FD can claim its ownership anytime," lamented Prioranjan Tanchangya, headman of Goinchhari. "The people may get rights if the area now barren is de-reserved. Then they can have an address. The Farua people can never be evicted, this is a reality," said Joysen Tanchangya.

Sudatta Bikash Tanchangya, the general secretary of the Movement for the Protection of Forest and Land Rights in the CHT thinks one with no rights also does not have any liability. So, for the sake of protection of the forest it is imperative to ensure rights to land for the Farua people. "The ethnic people of Farua are scattered. They should be brought together in a designated area and that area can be dereserved. This will ensure the rights of the people in the reserved land," said Sudatta Bikash Tanchangya.

Barrister Raja Devasish Roy, chief of the Chakma Circle, a leading Adivasi rights activist, and a researcher on the CHT land and forests shared his experience of a time (2007-2008) when he served as a special assistant to chief of the caretaker government: "Once I requested the Forest Department officials to show forest in the reserved forest. None of them could show me such forest. Actually, forest does not exist in any reserved forest except for some parts of Sangu Reserved Forest. Some forests still exist only in the *mouza* forest. In principle, I have no disagreement with the demand of the Farua people for dereservation. But if the state ensures the legitimate rights of the Farua people as citizens, the problem can still be solved without dereservation. He gave an example that, "To establish a school, the requirement is only 30 decimals of land, not 100 acres." He thinks a humanist approach is required to solve the problems of the Farua people to avoid the complexity of laws.

Reference:

Gain, Philip. 2000. *Life and Nature at Risk* in *The Chittagong Hill Tracts: Life and Nature at Risk*. Society for Environment and Human Development (SEHD), Dhaka.

Roy, Devasish. Forests, Forestry and Indigenous People in The Chittagong Hill Tracts (CHT). *Earth Touch*. No. 3, February 1997, SEHD, Dhaka.

Ek Najare Rangamati Circle (The Rangamati Circle at a Glance). 2011 (unpublished).

Ishaq, Muhammad (ed.). 1971. Bangladesh District Gazetteer: Chittagong Hill Tracts.

Dey, Tapan Kumar. 1999. *Bon Bebosthaponai Parbotto Chatagramer Dakshin Bon Bibhag* (The CHT South Forest Division in Forest Management). The CHT South Forest Division, Rangamati.

A forest village in Reingkhyong Reserved Forest.   © Philip Gain

Kassalong Reserved Forest

Sayam U. Chowdhury

Kassalong Reserved Forest.   © Sayam U. Chowdhury

While most of Bangladesh's semi evergreen forests in the northeast are on their last legs, forest patches in the Chittagong Hill Tracts in southeastern Bangladesh, especially Kassalong Reserve Forest remained ecologically uncharted. Kassalong was declared a reserve forest in 1881, yet very little is known about its biological diversity and it will probably disappear before its secrets are revealed.

In the Chittagong Hill Tracts, the Kassalong Reserve Forest is located bordering Myanmar to the south and east, with Assam and Tripura in India to the east and north and to the west by the Chittagong District and the extension of Hindu Kush-Himalayas region. According to statistics of the Forest Department 117,000 ha land was covered by natural forests and bamboos in 1963, which shrank to 73,000 ha in 1983 and further reduced to 65,000 ha in 1992. This shows approximately 10% decline in the forest cover every 10 years. Thus, based on Forest Department's record more than 50% forest cover in Kassalong has been reduced to date since 1963.

However, in reality the destruction was even more massive, leaving only less than 20% of the natural habitat in the last 50 years. The centenary of deforestation took its peak after the Chittagong Hill Tracts Peace Treaty was signed in 1997.

The indigenous peoples of the Chittagong Hill Tracts formed an armed force "Shanti Bahini" to protect their rights for land, life, and livelihood. The bush war came to an end in 1997 through a peace accord. The government assured hill dwellers to return to their niche in exchange of their weapons. This event completely altered the fate of the virgin Hill Tracts, which is largely correlated to severe damage of the natural habitats. The Shanti Bahini were living a semi-nomadic life hiding in the bushes, taking cover under dense natural forest, using the nature of their land as a mode of their survival. They became the guardians of nature for their own good and coexisted quite well with the forest and other forest dependent creatures unlike the plains-landers.

Some might blame slash-and-burn or *jum* cultivation in the Chittagong Hill Tracts as the primary reason for environmental and ecological degradation. However, politics, unsustainable development, and several other notable reasons also created an interconnected complex situation. For example, the Kaptai Hydroelectricity Project inundated 54 thousand acres of land, which pushed a huge percentage of hill people to the higher land or even outside of Bangladesh border. This reduction in fertile cultivable land significantly increased *jum* cultivation. Moreover, the decrease in land per person led to faster shifting cultivation. In the traditional practice a *jumia* would normally abandon a piece of land for 15 to 20 years to allow regrowth its fertility. However, this system of *jum* cultivation has greatly reduced lately. Thus, the Kaptai dam heavily impacted both the socio-economic and environmental condition of the Chittagong Hill Tract.

As the Shanti Bahini left the pristine forests, greed grasped the green land, making it a lifeless garden. The dwindling forest of Kassalong outlines a strong example of that. A road was built through the heart of the Kassalong Reserve Forest in 2003 starting from Baghaihat via Masalong and ended at Sajek, opening up a gateway to a mass devastation of the hill forests and its residents. The most interesting fact behind this immense logging is that the chopped up woods and timber from Kassalong travelled through only one route where the Bangladesh Army, the Border Guard Bangladesh (BGB), the Policy and Forest Department hawk day and night. To pass through many of these posts along the road the locals would normally face serious security checks and no vehicle can go in or out after dusk. This high-level-security takes it peak when a tourist or a journalist comes to these check posts – apart from passport and visa every possible way of identification must be shown to enter into the hilly terrain. It is awfully surprising how the million dollar timbers of Kassalong Reserve Forest did pass through the unremitting check posts of several parties of the Government of Bangladesh!

The aroma of fresh woods and illegitimate deals with law enforcement agencies attracted Bangali timber traders to Kassalong Reserve Forest, who hired the deprived locals to wipe out the reserve forest and fill up their trucks with heavy timbers.

The current insurgency groups along with the wildlife are taking shelter in the last remaining patches of natural forest. The remnants and leftover bits of the Kassalong Reserve Forest still support numerous rare and globally threatened wild animals. Globally *Endangered* mammals like Asiatic Wild Dog or Dhole (only 2,500 mature individuals are left in the world), Asian Elephant and Hog Deer are still fighting their last battle for survival in the fringes of Kassalong Reserve Forest. Moreover, globally vulnerable Binturong, Asian Black Bear, Gaur or Indian Bison and Sambar Deer occur in the fragments of the reserve. It is imperative to save the last vestiges of this forest of global significance.

Is it too late for interventions to preserve whatever is left of Kassalong? Are there measures to be taken to put an end to the destruction? Of course there is room for that, but without pointing any fingers, it is important the right authorities, the government, the timber businessmen, and the locals realize what they have caused and how far this can go before no trees are left to cut or give shade. Kassalong is just an example of what might be happening to many other reserved forests in Bangladesh. The first steps come with the construction of roads, then showing bias and misusing policies, then the businessmen arrive to show profit and with that Bangladesh grows less and less short of the right kind of green.

But one must remain idealistic and think it is not too late to conserve the remaining forest in Kassalong and with it preserve the biodiversity. The government and other conservation focused agencies need to step in to start restoration of the destroyed forest. It is not an easy task but it is not impossible. Bangladesh has seen enough damage done to its resources even prior to and post 1971 and now 41 years after liberation. It is time to stop the selfish agenda, cutting, and destroying and to initiate recreation and restoration what was once already there, for at least the chance to have a greener tomorrow.

Sangu and Matamuhuri Reserved Forests in Peril

Buddhajyoti Chakma

The Sangu and Matamuhuri reserved forests in the Chittagong Hill Tracts (CHT) are now severely degraded, to the point of extinction. The British colonial authorities created the Matamuhuri Reserve forest in 1880 and the Sangu Reserved Forest in 1881. The Matamuri Reserved forest is now believed to be 90% destroyed. However, the local *Adivasis* believe that if the invasion of marauders is completely halted, these two reserved forests will revive within 10-15 years. The major factors that have led to the destruction of these two reserves include unplanned development, emerging human habitation, pillage of wood and other forest produces, corruption of the Forest Department (FD), illegal logging, government sanction for logging, grabbing of land for rubber and horticulture, militarization, and the common man's helplessness in front of the influential people and groups. The fear is if these threats are not contained, the reserves may go extinct.

These two reserved forests of 184,000 acres are adjacent to the Arakan Hills, which are also the source of Sangu (locally known as *Shrinkho*) River and Matamuhuri River. The Forest Department (FD) and forest villagers (residents of the FD's villages) inform that all of 82,000 acres (33,184.22 ha) of the Sangu reserved Forest is natural forest. Of the 102,000 acres (41,277.94 ha) of the Matamuhuri reserve, only 2,000 acres had been cut for commercial plantation in 1952. The communication inaccessibility had shielded the Matamuhuri Reserved Forest from the greed of influential people. The forest remained almost intact till 1990.

After the democratic government came to power in 1991, the politically influential groups started to cut the forest indiscriminately. Groups with political backing brought in thousands of laborers to cut the forest from 1996 to 2001. At Alikadam Upazila Town, hundreds of thousands cubic feet of wood had been openly smuggled in hundreds of trucks along the Alikadam-Lama-Chokoria road right in front of the security forces and check posts of the security forces and the FD.

A few separatist rebel groups of Myanmar reportedly took shelter in the deep Sangu Reserved Forest from the 1980s. Later on, some families of Boro and Choto Madak, Remakri, Tindu, and Thanchi adjacent to the reserve started inhabiting there as well. They engaged in *jum* and poppy cultivation, causing some destruction of the forest. They were evicted from the area in 2007 and brought to the Boro and Choto Madak regions. The local people think that if adequate measures are taken to restrict the entry and human habitation in the reserved forest as specified by law, the natural forest can be restored in the next 10 to 15 years. But the FD officials fear that the government plans to construct roads from Tindu via Remakri and Boro Modok to Laikre in the Sangu Reserved Forest and from Alikadam via Babupara and Janalipara to Krokpatajhiri in the Matamuhuri Reserved Forest, which will cause rapid demise of these two reserved forests.

It is due to the lack of communication infrastructure and security that the Sangu Reserved Forest remained impenetrable to the FD workers since its was created. In the Matamuhuri Reserved Forest, the FD officials started sporadic patrolling of the areas adjacent to Alikadam upazila town from 1950. In 1952, it planted trees of many species including Garjan and Champa flower in almost 2,000 acres in Babupara, Roambhu and Janalipara areas. In some areas of Babupara and Roambhu, villagers from outside were brought in. Ukkajai Chakma, the villager headman (not a *mouza* headman, but a headman in reserved forest with limited jurisdiction) of Babupara, informed that the forest bandits started arriving ever since the Alikadam-Lama-Chokoria road was constructed in the 1980s. However, insurgency war between the Shanti Bahini and the military prevented the forest bandits from smuggling wood. Some internally displaced Adivasi families, who were victims of the unrest, took shelter in the deep of the reserved forest. Although their *jum* cultivation caused some harm to the forest, the mother trees remained unharmed.

The forest remained mostly intact till 1990. Ukkajai Chakma informed that unfortunately in 15 years (from 1991 and 2005), forest bandits protected by the political leaders, destroyed 90% of the forest. He said when the bandits engaged thousands of laborers and openly smuggled wood, the FD and the villagers had no other option but to silently observe the destruction of the forest as it was carried out with the consent of the politically influential people at the highest level of the government. Thus, protest never brought any results; it rather caused harassment to the protesters. He also informs that only 10% of the forest is left in the unreachable areas. However, the Lama Divisional Forest Officer, Harunu-ur-Rashid disagreed that 90% of forest was destroyed. In his assumption, 70% of the forest still survives full of trees. He thinks that the initiatives of the forest department alone are not enough to save the forest. Mr. Harun emphasized the necessity of local intervention as well as political goodwill.

The three forest divisions (Bandarban, Lama, and Pulpwood with a portion of Rajasthali Upazila of Rangamati Hill District) in Bandarban Hill District controls 264,524.20 acres of reserved and protected forests. Although the Deputy Commissioner controls *jum* land in the USF, the FD controls 38,644.75 acres of land under the process of being declared as reserved forest. An academic paper of Assistant Conservator of Forest (ACF) of Bandarban, Md. Eusouf divides the land of Bandarban into four classes according to use. These are: (a) uncultivated land— encompassing 100,945 ha or 6% of the total area (occupied by human inhabitance, roads, and rivers), (b) cultivated land—encompassing 240,473 ha or 20% of the total area (used for cultivation), (c) FD—38% of the total land, and (d) fallow land—36% of the total land. Although the FD alone controls the major share i.e. 38% of the land, in reality, there is no forest on most of these lands. However, on paper, these are reserved forests. Under these circumstances, the ACF Eusouf mentions that the decline in trees in the government and privately owned forests is now putting severe pressure on the Sangu and Matamuhuri reserved forests. ZuamLian Amlai, the Bandarban district president of Movement for the Protection for Land Rights and Forests in the CHT, fears that none can save the reserved forests unless the government shows strong will and commitment, and cancels the road construction project that runs through the forests.

Unplanned construction of roads (two pending approval as of March 2013) poses particular threat to the reserved forest. This causes loss of the land for the *Adivasis* and destroys the forest adjacent to the roads. Threatened by the intricate problems, the *Adivasis* move deeper into the forest to inhabit and to continue with *jum* cultivation. This adversely affects the reserved forests. There are no study and analysis of the risks and the mitigation measures to be taken if roads and other infrastructure are developed. As a result the forest and *Adivasis* are both falling victims to the crisis. The political leaders and influential people have recently contributed significantly to the destruction of the forests. The Matamuhuri Reserved Forest survived two decades of unrest (insurgency) in the CHT and military rules since liberation. But after the democratic government came to power 1991, the political leaders got actively involved in the plunder of the forest. The Ministry of Environment and Forests was allegedly involved in the pillage of the forests. The intrusion and inhabitation of outsiders is also a big threat. According to the forest law, it is illegal to enter the reserved forest without the permission of the FD, let alone inhabiting there. But ignoring this law, human habitation started in the Sangu and Matamuhuri reserved forests from the 80s. In 2007, the police and BDR (now BGB) evacuated more than two hundred families from the Sangu Reserved Forest and brought them to the Boro and Choto Madak area. The FD people have admitted that in 2012 some 12,000 people were living in Janali Para, Croakpata, Puamuri, and Indu-Sindu areas of Matamuhuri Reserve Forest. These people now reportedly demand a separate Union Council union with road communications.

The FD, according to Md. Eusouf, prepared a work plan between 1958 and 1962 for proper management of these two reserved forests. At that time 160 species of trees, 10 species of bamboos, 24 species of shrubs, 23 species of medicinal plants, nine species of Palmyra, 52

Northern Water Skink, one of the rarest skinks of the world found at Sangu-Matamuhuri Reserved Forest.   © Sayam U. Chowdhury

species of climber, 18 species of reeds and grass, and 14 species of parasitic plants were detected. But the plan does not mention anything about wild animals. The elderly people inform that in the past hundreds of species of wildlife and birds including tiger, bear, varieties of cats, different types of deer, elephants, large dogs, large goats, monkeys, entellus, gibbons, caped langur, bumble bees, peacocks, wild cocks, kaleej pheasants, hornbills, parrots, and mynahs. Now most of these animals and plants have become extinct or reduced to a miserable number due to the vanishing forest and the emerging human habitation. Many of the animals have migrated to the neighboring Myanmar's deep forest.

The plan developed by a Deputy Conservator of Forest, Bahauddin Chowdhury, divided the two reserved forests into three circles, 63 blocks, and 105 compartments. The three circles, divided on the basis of land classification and their suitability to produce trees and bamboos are: (a) hard wood work plan (17097.97 ha in Sangu and 16662.53 ha in Matamuhuri), (b) pulpwood (8,149.15 ha in Sangu and 8,522.27 ha in Matamuhuri), and (c) bamboo (32,976.58 ha in Sangu and 41,201.75 ha in Matamuhuri). The Sangu Reserved Forest is divided into 32 blocks and 53 compartments while the Matamuhuri Reserved Forest is divided into 31 blocks and 52 compartments. Some initiatives were taken during the Pakistani regime in implementing the work plan, but the plan was discontinued after the liberation of Bangladesh.

Lama Divisional Officer, Harun-ur-Rashid admitted that the Matamuhuri Reserved Forest has been partially destroyed. Pointing to the current condition, he said that illegal human habitation is growing in the forest and the Local Government Engineering Department (LGED) has taken initiatives to create an establishment there. He also informed that a LGED instigated road construction project has been cancelled due to concerns regarding further forest destruction and human inhabitation.

The Divisional Forest Officer (DFO) of Bandarban, Tosibul Bari Khan said that the Sangu Reserved Forest is home to the only virgin forest in the country. The DFO said that although the *jum* cultivation by the *Adivasis* has caused the wearing out of 30% of the forest, the rest of the 70% still remains beyond human intrusion. Mr. Khan said that all measures would be taken to protect the forest including the prevention of unplanned road construction.

Valuing Village Commons in Forestry: A Case from the Chittagong Hill Tracts

Raja Devasish Roy and Sadeka Halim

Introduction

The village common forests ("VCFs") in the Chittagong Hill Tracts (CHT), in the wider sense, includes any forested area that is used by village communities on a collective basis and which is regarded as their common property, irrespective of its legal classification. In a narrower sense and within the context of this article, VCFs will refer only to such forested areas of the CHT that are situated within the boundaries of *mauzas* and directly managed, protected and used by indigenous village communities. VCFs in this narrow sense are also known as "mauza reserves" or "service" forests in the CHT and they are located outside of the reserved forest areas that are administered by the Ministry of Environment and Forest. Therefore, VCFs in this sense will exclude forest commons within the reserved forests. Although there have been attempts to record the title of some VCFs in the name of Buddhist monastery committees, the vast majority of them are not recorded in the district land registry in the name of any legal entity. The only known exceptions are a 80-acre VCF in Barkal sub-district and a 250 acre VCF attached to a Buddhist monastery-cum-meditation centre on the border of Rangamati (Sadar) and Kaokhali sub-districts, within Rangamati district. In these two cases, the Rangamati hill district council, upon the recommendation and consent of the concerned mauza headmen, recognized the VCFs and resolved to not allow settlement, lease, acquisition or other transfer of the VCF lands to any but the organization that manages it. Rule 41A of the CHT Regulation of 1900 refers to these VCFs as "mauza reserves" but does not provide for any system of titling or registration.

Although the primary object of our discussion is related to these VCFs, it is of course not possible to obtain an in-depth understanding of the decline of the VCFs in isolation from the general trends in the decline of forest commons in the wider sense, both within the CHT and elsewhere in this sub-continent. Similarly, we have also found it necessary to relate this discussion within a broader framework that analyses the major trends in resource management patterns in the CHT within the context of local, national and regional resource use policies and related social, political and economic developments.

Historical Background: State Appropriation of the Forest Commons

Forest Commons in Pre-British India: The existence of forest commons in the wider sense in the Indian subcontinent can be traced back historically to the pre-colonial period. In pre-British South Asia, a system of local control evolved to conserve forested lands. Despite the inequalities of caste and class, pre-colonial society had a considerable degree of coherence and stability. When environmental degradation and deforestation started to occur in the wake of the early wave of sedentary settlements in this sub-continent, changes in social structures were devised in response to this "resource crunch". Even the Mughals did not dismantle the existing pattern of resource use and the social structures in which they were embedded. In pre-British South Asia, the control and management of local resources was vested in local communities who designed a variety of practices for effective conservation and sustainable use of forest resources (Gadgil and Guha, 1992:103-112; cited in Halim, 1999.a: 69).

Forest & Swidden Commons in pre-colonial CHT: Turning now to the Chittagong Hill Tracts (CHT) in southeastern Bangladesh, we see that this mountainous frontier region was densely forested even up to the last quarter of the 19th century. This prompted the British colonial government to declare 5,670 square miles out of the then total area of the CHT's 6,882 square miles "as government forest" in accordance with section 2 of Act VII of 1865 (Ishaq 1975:107). Prior to such declaration, there was no formal distinction between forest and other lands. The entire region of the CHT was then used by its eleven indigenous peoples for their homesteads, their swidden or *jum* plots and as a repository of natural resources for their domestic use, and to a limited extent, for trade with market settlements in the coastal region of Chittagong. These forested lands were rotationally cropped for swiddens or *jum* whereby each plot of cultivated land was left for fallow for several years. Therefore, although *juming* involves the clearing of the forest growth (except large trees), the land regenerates itself into a forest again within the fallow period, to be *jumed* again or to be left as a forest. Therefore, the factual distinction between *jum* lands and forest lands is merely a transitory phase depending upon the use of the land at any given time, Thus, in many cases, the same lands were both forest and swidden lands (Roy 1997.a: 3).

Converting Forests into Mono Plantations: Such a state of affairs did not last long. As mentioned earlier, soon after the annexation of the CHT in 1860, the British government came to regard four-fifths of the CHT as "forest". They regulated the use of these "forests", but their direct administration was left in the charge of the *Chiefs* or rajas and their subordinate indigenous officials. Direct government management was restricted to the newly-created category of "reserved forests" (RFs), whose total area in 1884 was 1,345 square miles (Ishaq 1975:109).[1] A large part of these forests of heterogeneous stand was clear-felled and converted into mono plantations of industrially and commercially valuable species like teak or *gamar* (Gmelina Arborea), neither of which are indigenous to the CHT. The Pakistani and Bangladeshi governments followed these trends and added plywood species, and even other imported species including, Acacia, Eucalyptus and pine (Gain 1998).

Appropriating the Forest Commons: It has been said that the "forest reserves" (RFs) in different parts of South Asia were established to restrict the "untrammeled" use of the forest resources by forest dwellers (Gadgil and Guha 1993). The British-promulgated the Forest Act of 1927. This is still valid law in Bangladesh and in India and based upon the earlier Indian Forest Act of 1878. The 1927 Act provided that no person could claim right to private property within reserved forests merely because he or she is domiciled there. With such a narrow interpretation of the laws, the forest dwellers could not have their habitats recognised as their property, common or private, with some notable exceptions in India in recent times.[2]

However, despite the limitations invoked by the colonial authorities, forest-dwellers continued to inhabit areas categorized as "forests" for as long as they could. Historically, forest dwellers have never actually owned or possessed forests in the modern legal sense. Thus the British administration invoked their "monarchical" claim over these hitherto commons and purported to exercise their near-absolute claims through

"Acts of State" (Halim 1999.a: 72-76, Roy & Halim 2001.a: 7-10,). Where there were chiefdoms that were oriented around a system of clans rather than clearly-demarcated geographical areas, as in the CHT, the British government acknowledged that the chiefs had some authority over their people, but none over the land, which was said to belong to the state alone. This process of appropriation has been described as an act of "legal gymnastics", which is uncannily reminiscent of the infamous *terra nullius* principle that was invoked to appropriate the ancestral lands of the Australian aborigines (Roy 1994:13). As in the CHT, the British government continued to confiscate forests in other parts of South Asia throughout the second half of the nineteenth century (1850-1900) (Grove 1995; cited in Halim 1999.a.).

Unidimensional Scientific Forestry Rejects Indigenous Forest Management Practices: This marked the beginning of what is called "scientific management" of forests, signifying a more intense, but largely unsustainable, pattern of use of forest resources in this sub-continent. The major objective of such manner of management was the maximization of production and/or harvesting of commercially valuable timber and other "minor" forest produce (like bamboo, rattan, cane, etc) while ignoring the ecological, medical, cultural and micro-economic needs of the local communities (Halim 1999.a.: 73-76, Halim 2001.a: 14). These policies were executed by the newly-created Forest Department that functioned in the nature of a police department at times—with armed guards who patrolled the forests—and an administrative department at others. Although some planting was done, they were restricted to mono plantations.

The net result of the aforesaid measures are accounted to have been the major cause behind the erosion of traditional rights of forest-dwelling peoples and the erosion of the rich tradition of forest conservation in this subcontinent. The perception of forest ecosystems as having multiple functions for satisfying diverse and vital human needs for air, water, and food was superseded by the growth of the uni-dimensional "scientific" forestry.

Formally, the Forest policies did recognize, to some extent, the importance of conservation and protection, but the importance of forest commons to the rural poor, although grudgingly accounted for, was never promoted actively. Thus, although the Forest Act of 1927 provides–through section 28–for the assignment by the government of land and resource rights within reserved forests to specific village communities (which were hitherto their commons), this provision has never been acted upon to any significant extent, or at all, by the British government, and even less so by the successor governments in India, Pakistan (including the then East Pakistan) and Bangladesh. As Scott (1998) has pointed out, this was because the state saw commons as "un-exploited" and "fiscally barren".

VCFs: An Indigenous Innovation

With the beginning of British rule in the CHT in 1860, the indigenous people not only lost their right of access to a quarter of the entire area of the region (which came to be categorised as reserved forests), but at about the same time, large forested areas were also converted into plough lands, which yielded high taxes to the revenue-hungry colonial government. Further reduction of access to forests was to follow in the wake of the Kaptai Dam in the 1960s and the population transfer or "transmigration" programme of the 1970s.

To return to the British period, the indigenous villagers who lost access to their former commons now found themselves with little choice but to devise new methods of sustainable use of their now-scarce common lands. The result was an innovation based upon their traditional resource management patterns to retain forest cover for long-term use. This gave birth to the village common forests (VCFs) of today, which are not allowed to be cultivated for *jum* or otherwise by the communities themselves, on the strength of sanctions and religious taboos. The maintenance of VCFs was combined with approaches to prevent a shortening of the fallow periods on lands that were left outside of the VCFs. The latter proved to be even more difficult due to population rise and other causes, as we shall discuss in more detail below. This in turn also affected the efficacy of the VCF-protection measures, along with other constraints that were now faced by CHT villagers.

The Village Common Forests of Today[3]: The current pattern of VCF management in the CHT involves only semi-structured or unstructured methods. In some cases, VCF management involves the entire adult population of a particular village. In some cases, village communities have formed unincorporated associations with restricted membership and elected office-bearers and even reduced their use and resource-sharing practices into formalised rules. However, the unstructured models are more common, and these are usually centred round the leadership of the *karbaries* (singular: karbari) and the *mauza headmen*. These headmen in the more than 350 *mauzas* of the CHT are nominated by the chiefs or *rajas*, but formally appointed by the deputy commissioners (Roy & Halim, 2006). The far more numerous *karbaries* are traditionally nominated by the villagers and appointed by the chiefs. These offices have now become largely hereditary and are confined to men only, except to a limited extent in the case of headmen.

The aforesaid innovative practices did not escape the astute notice of the British administrators. Although, on the one hand, the British continued to deny access of the indigenous people to the reserved forests, they realised that unless these village forests were protected, villagers may have no recourse but to permanently migrate into or otherwise utilise the resources of the reserved forests, which had come to be regarded as the exclusive property of the state to the exclusion of indigenous peoples and other forest dwellers. Thus, through an amendment to rule 41 of the CHT Regulation of 1900 in July 1939, the government recognised a category of "*mauza* reserve" that was to be identified, demarcated, and protected by the *mauza* headmen. Villagers have come to refer to these forests as "service" forests, since they serve their village community.

These VCFs are now under severe threat due to a variety of factors including population rise and the consequent growth of village settlements, the spread of sedentary agriculture, horticulture and tree plantations, and in-migration and out-migration. A more detailed discussion of these causes is contained in the section following the discussion on the multidimensional importance of VCFs to the CHT village communities.

The Ecological, Economic, and Cultural Significance of VCFs

The exact number and extent of VCFs within the CHT are not known, as there is as yet no organised database upon it, but they are certainly known to exist in all three districts of the CHT. They are perhaps more numerous in Rangamati than in the other two hill districts. Most of the reported VCFs are small in area, ranging from about 20 acres to more than 400 acres. Some of the larger VCFs are located within Jurochari, Barkal and Langadu *upazilas* (sub-districts) within Rangamati district. Some consist predominantly of bamboo brakes, while others contain a more heterogeneous stand of flora, and consequently, fauna as well. Many VCFs also contain the herbaria for the village concerned, which the local *viadya* or *ojhas* use to prepare their traditional medicine, while some are regarded as sacred. Many Buddhist monasteries within the CHT are situated within the middle of a forest or small woods. Thus the maintenance of VCFs is also crucial to safeguard the cultural integrity of these peoples.

The larger VCFs also contain natural springs and other aquifers that are used for drinking water. With the consent of the VCF community, individual families may extract wood and other natural resources for their domestic use. Sometimes, where it can be done sustainably, villagers also sell some of the forest produce, usually bamboo and less occasionally, timber, to meet community needs for school and temple construction and for emergency medical expenses. Villagers insist that with some species of bamboo, occasional harvesting actually leads to further growth rather than impede growth. The economic importance of forest commons in meeting essential biomass needs of rural communities has been emphasised by many writers such as Agarwal (1989), Ravindranath (1996), and Gadgil and Campbell (1996) (cited in Halim 1999.a: 58).

Thus VCFs are of immense value for environmental, economic, medical and cultural reasons. This is especially so since natural resources in the CHT are becoming more and more scarce and negative ecological changes have driven the hill people into severe hardship. The main factors why they are so important are:

■ Given the rate of the deforestation in Bangladesh, (more than 3.3% annually in Bangladesh; Forestry Master Plan 1993: 2) it is imperative for environmental reasons to maintain and protect these forests.

■ VCFs are also huge repositories of biodiversity. They are the homes of diverse animal and plant life, including herbs and plants used in indigenous medicine, which have a significant potential for modern medical science.

■ The VCFs are the main sources of wood and bamboo required for house-building, medicinal and other sustainable biomass needs of hill villagers.

■ The traditional use of VCF produced by hill people (like, medicinal herbs, fuel wood and fodder) has kept pressure off the government-owned reserved and protected forests.

■ VCFs are crucial for watershed management. Many VCFs contain natural springs, and headwaters of streams and other aquifers.

- VCFs are also related to the spiritual and cultural beliefs, rituals, and ceremonies of many indigenous peoples.

Understanding the Causes of the Decline of VCFs: The indigenous peoples of the CHT had started to lose control over and access to many of their forests due to the Forest policies of the British colonial government. These policies were followed during the Pakistan period (1947-1971) and reinforced by the successive Bangladeshi governments from 1971 to this day, barring a few exceptions that suggest a small trend towards a more "participatory" approach to forest management. It is important to revisit the Forest policies from the British period (1860-1947) in conjunction with other land and resource management policies and practices until today and relate them to broader political and economic developments that have directly or indirectly affected the fate of the VCFs.

Forest Laws and Policies

It may be recalled that the Forest laws and policies of the British period had rejected community management of forests in favour of forest management through a government agency, namely, the Forest Department, which functioned in the nature of a police department-cum-administrative department. This emphasis was not to change, and although "forestry" programmes of a more participatory nature were also implemented from the 1980s onwards, the Forest Department until today has failed or refused to concentrate its efforts to provide extension services to village communities or homestead foresters. Even unto this century, the Forest Department considers that the categorisation of forestlands into "Reserved Forests" is the most efficient way to manage and protect forests. From the 1990s and even up to 2000, the government has continued to expand its area of reserved forests within the CHT, a process that has been vehemently resisted by indigenous farmers who rallied round a mass organisation known as the Movement for the Protection of Forest and Land Rights in the CHT (Roy, C.K. 2000:178-180).

Land Laws & Policies

Unlike in the case of Forest-related laws, policies on the ownership and use of CHT lands other than forests were to see significant shifts, starting from the British period to the Pakistani period and post-independence

Bangladesh. Private leases for plough lands started to be recognised in the first quarter of the 19th century. Private ownership of hillside lands, however, was to be recognised only since the 1950s. However, a more drastic change, at least in the eyes of the indigenous people, was to come with the opening up of land ownership within the CHT to non-resident individuals and corporate bodies. An amendment to Rule 34 of the CHT Regulation in 1971 and then again in 1979 (the 1979 law is almost a ditto copy of the 1971 law) reduced the area of unclaimed public land that could be settled or leased out to local farmers from 25 acres to 5-10 acres. At the same time, the new law allowed non-residents to acquire land rights within the CHT for homesteads, commercial plantations, and industrial plants. In the case of the latter, leases for hundreds of acres could now be obtained (by non-residents) without the knowledge and consent of the chiefs and headmen, which was hitherto near impossible. This was contrary to the letter and spirit of the CHT Regulation, which regarded the CHT primarily as a homeland for its indigenous peoples, whose primacy with regard to land and resource rights was guaranteed as against outsiders. Since the CHT Regulation was bereft of constitutional protection since 1964, there were no concerted attempts to challenge the constitutional validity of these laws.

Changes in Resource Use & Livelihood Patterns

The aforesaid laws were implemented through a broad range of government-managed programmes and projects resulting in high in-migration and promotion of market-oriented horticulture and tree plantations that led to the conversion of many VCFs into orchards and plantations. However, the projects on horticulture by the Forest Department's Jum Control Division and the CHT Development Board proved to be almost total failures (Sattar 1995:11, Roy 1998: 95). In conjunction with the growing integration of the CHT economy with the market economy of the plains, these policy shifts were to result in fundamental occupational changes among the region's farming population even outside of formal government programmes and projects.

From a broad perspective, it is difficult not to reach the conclusion that the decline and degradation of forests in general, and VCFs in particular, is directly related to the scarcity of land and other natural resources vis-à-vis the constantly growing population, which subsists upon them. Apart from the appropriation of the reserved forests, some major occupational changes among the indigenous population also led to a huge reduction of the forest cover of the CHT. These include the introduction of plough cultivation—largely wet-rice farming—in the first quarter of the 19th century, the growth of sedentary orchards for market-oriented horticulture in the 1960s and the constantly growing practice of tree farming by settled indigenous farmers in the 1970s (Roy 2000.c: 101-103). It may be noted that we include tree farming among the causes of deforestation since we do not wish to consider tree plantations with narrow genetic bases (predominantly teak and *Gamar or* Gmelina Arborea) as "forests" in the usual sense of the word. In the case of plough cultivation and horticulture, it was not only the cultivated patches that were deforested but also the surrounding areas, since wildlife from the forests were regarded as a threat both to the cultivated plot and to the safety of the farmers.

The Kaptai Dam and In-migration

Available resources are known to have already reached crisis proportions by the 1950s, well before the Kaptai Dam was built (Sopher 1963). A direct consequence of the creation of the Karnaphuli reservoir, which was created by the Kaptai Dam in 1960, was the (so far) permanent inundation of two-fifths of the entire ploughlands of the region and a large part of the Rangamati Reserve and other small reserved forests. Its indirect consequences were almost as severe. The estimated 8,000 *jumia* families of the affected area did not receive any compensation by way of money or alternate lands. Therefore, a large section of these farmers who could not become horticulturists had little choice but to migrate to India. Others began to migrate into the un-submerged uplands that were already populated. Many plough farmers also did likewise, forcing the people of the un-submerged areas to share their *jumlands* and VCFs with them. This led to a huge reduction of the fallow cycle in *jums* from about seven years in the 1950s to a mere three to four years in the 1960s (it was above ten years during the British period), seriously upsetting the ecological balance of these areas (Roy 1997.a, Roy).[4] Evacuees from within the Bilaichari upazila were forced to migrate into the Reingkhyong Reserved forest to seek a new livelihood as "forest villagers" of the Forest Department, essentially little more than modern day serfs. These environmental refugees are not allowed to acquire any land rights and are bereft of social extension services like basic education and healthcare (Roy & Gain 1999, Roy & Halim 2001.a, Wangza, 2011).

The population transfer programme of 1979-80s was to further exacerbate the problem. Tens of thousands of Bengali settler families (numbering anywhere between 250,000 to 400,000 people) were resettled on lands, including VCFs, that were owned and/or occupied by indigenous people (Roy 1997.b). These settlements led to the violation of the indigenous peoples' common resource rights by the newcomers. The encroachments over the lands of the indigenous people took place mainly because the migrants did not have sufficient cultivable land to sustain on (Roy 1998: 76-77). It has been argued that the land grants to the settlers were made in violation of the CHT Regulation and other customary and statute laws, but the matter has never been litigated in court (Roy 1997.b.). It remains to be seen how the *CHT Land Commission*— which was constituted in accordance with the "Peace" Accord of 1997—decides upon the issue when it inevitably comes before it. The potential conflict between state law and customary law is a cause for serious concern here (Roy, 2000.e.).

Jum **Cultivation***:* Of course there are many other factors that have influenced the decline of VCFs and other forests. Government sources have constantly blamed *jum* cultivation as the major cause of deforestation in the CHT. We cannot agree. No doubt *jum* cultivation on forests of primary growth in today's resource-scarce age is hardly sustainable. Nor can we deny that over-cropping of *jumlands* can and does lead to a net loss of vegetative cover. However, we ought to note that *jum* cultivation is nowadays largely restricted to secondary growth forests, which had been *jumed* over previously. If we are to prevent over-*juming*, we should try to offer viable alternative livelihoods to *jumias*, such as through aiding their forestry, plantation, and horticultural ventures. In any case, illegal logging for timber, and for firewood to feed brick-making kilns, has certainly done far more damage to forests than *juming* has (Roy 1996). There is little doubt that these activities are far more harmful than *juming* because they "are [motivated] by pure profit backed by an almost limitless demand in the market, [which] pales into insignificance the extent of deforestation caused by [*jum*] farmers ..." (Roy 1997.a.: 36).

Other Causes*:* Among other causes of deforestation are the infamous CHT Forest Transit Rules of 1973 that make it extremely difficult and expensive for local farmers to obtain the mandatory permit to extract and sell the trees grown in their recorded lands.[5] Ostensibly, these rules were framed to prevent theft and pilferage from the government-controlled forests and plantations, but their injudicious application has resulted in huge corruption and a major disincentive to the growth of homestead plantations. If these Rules and their application were simplified, they would definitely lead to an increase in tree cover, raise the incomes of local farmers, and hugely reduce the pressure on government-managed forests and plantations, and the remaining VCFs (Roy 1998: 102).

Even if we hold that *juming* is partly responsible for deforestatation, we cannot excuse the government's role in pursuing short-sighted policies on resource use, including the Kaptai Dam, population transfer, the enhancement of the area of the reserved forests, and inequitable land distribution, that has deepened the resource crisis by the hugely shortened *jum* cycle. Therefore, it would be far more equitable and practical to address the underlying causes of deforestation—namely, resource scarcity and illegal trade in forest produce—rather than to blame its symptoms and minor causes, including *jum* cultivation. (Ibid.)(Roy 1996).

Adverse Impacts of the Decline of VCFs and Other Forests

The decline of forests in general, and VCFs in particular, had wide-ranging effects on the local environment and on the rural population of the CHT, whose economy is predominantly agrarian and dependent upon natural forest resources. Of course, sedentary and intensive farming did result in growing prosperity for some local people, but they were a small minority. When we take into account the irreparable harm that these developments caused to the local ecology and environment and the violent socio-cultural changes that they brought in their wake, the positive economic impact does not seem so significant. Rural women were to bear the brunt of these changes, as is explained below.

Environmental Damage: Among the most visible signs of environmental degradation that was caused by deforestation was the silting up of rivers, making navigation more and more difficult. Similarly, many springs, pools and other aquifurs that were used for drinking and for other domestic use began to dry up during the dry season, and sometimes even permanently. Many settlements had to be abandoned, including collective farm or *joutha* *khamar* sites of the CHT Development Board, as these resettled *jumia* farmers had to go farther and farther to fetch water. The decline in plant life led to a decreasing animal population, which in turn affected plant life since the consumption patterns of animals, who depended on other animals and/or plants for food, was also disrupted. This has set into train a complicated cyclic changes of life forms that has caused inestimable and irreparable loss and damage to the hitherto biodiverse plant and animal life. Bison (*bos frontalis*), Indian Rhinoceros and the two-horned Rhinoceros and many other species are now either extinct or nearly so (Ishaq 1975:13). The loss to flora is perhaps even worse, but it is difficult to estimate the nature and extent of this loss since there is no reliable data on these species, which are, or were, known only to indigenous people, and had not been "discovered" by botanists.

Adverse Impacts on Health: Loss of Food, Medicine & a Healthy Environment: The decline of animal and plant life, which formed an important dietary supplement to the fruits and vegetables grown by indigenous farmers, led to nutritional deficiencies. Since the VCFs and other forests also served as repositories of plants needed for indigenous medicine, traditional healthcare too suffered. The forests also contained bamboo and sun grass, needed for the indigenous people's houses, which were traditionally built on raised platforms for health and hygienic purposes, and also to stay beyond the easy reach of wild animals and mosquitoes. Deforestation led to housing materials becoming very scarce and expensive. It is therefore no wonder that one does not see these *machan* houses other than in the forest settlements in the relatively "remote" areas. Such houses are generally clean, airy, dry and do not require hillsides to be levelled. The space beneath the floor of these houses can also be utilised to house poultry and other domesticated animals. With the changing pattern of houses, landslides are now more common, and it is also believed that the growing scourges of malaria, typhoid and lung diseases are directly related to the relatively poorer housing conditions in comparison to the *machan* houses.

The Decline & Disappearance of Traditional Occupations[6]: With the decline of the naturally maintained herbaria, village shamans (*ojhjas* and *vaidyas*) could no longer ply their trade. Similarly, hunting used to be a secondary occupation of many indigenous farmers, and this could no longer be done. The Chakma word "Pollan", denoting a hunter, is rarely heard nowadays. Similarly, basket-weaving, which requires cane and bamboo, has also been affected. Thus, deforestation has also caused fundamental changes to the cultural traditions of many indigenous communities.

Social Impacts: Among the social impacts of deforestation are those that are related to out-migration and changes in occupational patterns. This has had major implications on class and gender through the disruption of social support networks. Social relationships with one's kith and kin, and with villagers outside the kin network provide economic, social, and political support that is important to all rural households. Deforestation has heightened social cleavages and sharpened the economic disparity among the indigenous population. This is because those who could acquire private property managed to prosper relative to their land-less neighbours who became poorer and poorer. There is no doubt that the brunt of the adverse impacts was borne by the poorer sections of the rural population, and by their women. Moreover, for forest-based communities, the relationship with forests is not only functional but also symbolic, embedded in cultural meanings. The decline of these common lands has eroded a whole way of living and thinking.

Impact on Indigenous Women: Researchers have pointed out that in developing countries it is women who are the most dependent on forests for their sustenance (Shiva 1989:18). The traditional division of labour in forest-dependent societies has allocated hazardous tasks as well as those requiring physical strength to men, and work that requires sustained effort and endurance has been assigned to women. The division is strengthened with taboos and beliefs. Deforestation affects indigenous women more than indigenous men because women's primary responsibilities such as cooking, fetching water, and gathering firewood pose hardships when ecological degradation of forests occurs. This is equally true in the CHT, especially for women from the most underprivileged section who have no private lands and are therefore highly dependent upon forests for their livelihood. However, rural women from the middle and higher income classes are also dependent to a large extent upon forest resources where the economy is at least partially subsistence-oriented and where wage labour is scarce both for economic and social reasons (hill people are generally averse to doing domestic labour for others). Thus the problems faced by rural women are closely related to environmental problems.

Recognition of Indigenous Knowledge & Practices: Since indigenous peoples have proved themselves to be efficient managers and custodians of forests, it is only natural that their concepts on forest management be given due recognition and application, as appropriate. Indigenous knowledge has been recognised as "traditional scientific knowledge" in Agenda 21, which was adopted at UNCED in Rio. The Convention on Biological Diversity, which resulted from the Rio process, also acknowledges the importance of the "knowledge, innovations and practices" of indigenous peoples related to the conservation and sustainable use of natural resources. Likewise, the knowledge systems of the CHT indigenous peoples are also worthy of protection and equitable utilisation, with their free, prior and informed consent, if so provided. It is unfortunate that it has never been formally acknowledged that the *taungya* method of raising plantations is a people's innovation based upon *jum* methods. Forest departments in Bangladesh and elsewhere and the UN agency, FAO continue to use this indigenous technology without either recognising the indigenous roots of this innovation or sharing its benefits with indigenous peoples, in violation of the Convention on Biological Diversity.

Security of Tenure: As in the case of indigenous knowledge related to resource use, a most important element in the protection of VCFs is the tenurial security of these lands for the communities concerned. Indigenous forest management perspectives on the tenurial status of forests have differed radically from the conventional industrial-capitalist concepts influenced by colonial legislative regimes. Indigenous communities living within and around the forests view themselves as keepers of forest heritage, which is passed down through the generations. These concepts need to be accounted for and acknowledged to ensure the sustainability of the VCFs. Therefore, a crucial need for policy makers is to recognise the significance of VCFs and the age-old formal and informal resource rights of the village communities living within and around the forests, and the knowledge systems related thereto.

At an international meeting related to the Convention on Biological Diversity that was held in Bonn, Germany in October 2001, indigenous participants expressed their strong reservations against the inequitable utilisation and marketisation of their forest and other common resources and the commodification and privatisation of their traditional knowledge systems related to natural resource management.[7] Indigenous communities in the CHT and in other parts of Bangladesh have a long legacy of managing and protecting VCFs. Over the centuries, many of these VCFs have been appropriated by the government under forest settlements acts that were negotiated in the late nineteenth or early twentieth centuries. The Bangladesh government took an active part in the Earth Summit and shifted towards people–oriented forestry programmes. This paper, therefore, strongly recommends the equitable redistribution of forestlands and the equitable utilisation of these resources and the related indigenous knowledge, based upon equitable principles.

The CHT Accord of 1997: Despite its various shortcomings, the CHT Accord of 1997 provides a reasonable basis upon which some of the aforesaid issues can be reasonably addressed, if not redressed in whole. Apart from recognising the legislative prerogative of the CHT councils, the Accord and subsequent legislation provide for two important safeguards for the indigenous people and other residents of the CHT, which, however, are yet to be acted upon in practice. One of these is the devolution of land administration to the hill district councils, without whose consent no lands are to be settled, leased out, transferred or compulsorily acquired (section 64, Hill District Council Acts, 1989). The other is the resolution of land-related disputes by a commission on land that is required to adjudicate in accordance with the "laws, practices and usages of the CHT" (CHT Land Commission Act, 2000). It is well to note that, if and when implemented, these processes could help restore dispossessed VCFs and other lands and prevent the privatisation of VCFs. However, we feel that both processes will not be enough to restore dispossessed lands or to prevent the privatisation of the VCFs. In addition, measures will also be necessary to prevent further deforestation in all the major types of forests through a range of measures, including incentives to local tree planters, joint-management of the reserved forests, giving registered title over VCFs to the concerned villagers, and so forth.

The implementation of the CHT Accord has however run into difficulties and there have been complaints that the pace of implementation is too slow (Roy 2000a). Moreover, land administration is yet to be devolved upon the hill district councils as stipulated in the 1997 Accord and subsequent legislation in 1998. The dysfunctionalities within the CHT administrative system, including the lack of cooperation between the CHT councils and line ministries in Dhaka also needs to be addressed (Roy 2000.b.).

The CHT Accord of 1997 has provided a sound basis to address many of the itinerant problems on resource use and ownership. The most important thing is to take this process forward in a meaningful manner so that the national and regional policies may be revised and reformulated to address the broader question of resource rights and deforestation, and the particular question of protecting and reviving the VCFs. Of course, a broad range of issues will need to be addressed. We suggest a number measures in particular. These are:

■ For the Hill District Councils to record the VCFs as the joint and common property of the concerned VCF communities;

■ For the Government of Bangladesh to redistribute state-appropriated common forest lands to indigenous communities conditional upon their sustainable use as forests (including by invoking section 28 of the Forest Act of 1927, framing rules under this section and issuing policy directives and necessary budgets) ;

■ For the Ministry of Environment & Forest to involve the indigenous peoples and other forest-dependent communities in the joint management of state-managed reserved and other forests and to share the resources of such forests in an equitable and practicable manner;

■ Aimed at the Government of Bangladesh to recognise the indigenous knowledge, innovations and practices related to forestry and environment protection and utlise them with the prior and informed consent of the peoples and communities concerned;

■ For the Government of Bangladesh to repeal amendments to the Forest Act of 1927 that were enacted in 2000;

■ For the Ministry of Environment to amend the Social Forestry Rules, in consultation with indigenous and other forest-dependent communities, and with their free, prior and informed consent;

■ For the Government of Bangladesh to cancel all notifications that purported to create new reserved forests in the CHT – from the late 1970s to 2000s – without the free, prior and informed consent of the concerned indigenous and other communities in violation of the letter and spirit of the CHT Accord of 1997;

■ For the Government of Bangladesh to repeal the Wildlife Act of 2012, which was framed in violation of the rights of indigenous and other forest-dependent communities; and

■ For the Government to refrain from amending the Forest Act of 1927 without the free, prior and informed consent of indigenous peoples and other forest-dependent communities.

However, we feel that most of the recommendations made above- except the first recommendation - can be realised only if they are complemented by other legislative and executive measures to recognise the resource and self-government rights of the indigenous peoples and to address the dysfunctionalities in the concerned administrative system, such as by implementing the recommendations contained in the *Rangamati Declaration of 1997* and the *Dhaka Forest Declaration of 2001.* These declarations contain many demands and suggestions that indigenous peoples and environmentalists have reached over the years on a consensual basis and are therefore worthy of serious attention if we really care about forests, biodiversity and basic human rights.

Notes

[1] According to Webb and Roberts (1976), the reserved forests covered almost 24% of the CHT (in 1976). From this we must deduct the areas in Langadu and Baghaichari upazilas of Rangamati, which were decategorised as reserved forests in 1979-80 to resettle government-sponsored Bengali transmigrants.

[2] Through the passage of the *Scheduled Tribes and Other Traditional* Forest Dwellers *(Recognition of Forest* Rights*) Act, 2006*, rights of forest-dwellers have been accorded some protection.

[3] The description of the CHT VCFs is based upon the authors' site visits and discussions with people related to the management of VCFs between 1996-2001 and upon data collected by the Rangamati-based NGO, *Taungya.*

[4] For a more detailed discussion, see Khisa, 1960:50, Ishaq, 1975: 88 and Haque, 1995:18 and Roy, 1997.a, to which studies the authors are indebted.

[5] Workshop on Environment, Forests, Forestry and Horticulture organised by Taungya in Rangamati on 5 September 1998.

[6] This section draws heavily upon Roy, 2000, which studied the occupational changes of the indigenous peoples from 1860 to 2000.

[7] Declaration coming forth from the *Sixth International Indigenous Forum on Biodiversity* (Bonn, Germany, 15-19 October 2001) which was presented at the *Ad Hoc Open-Ended Working Group on Access and Benefit Sharing*, Convention on Biological Diversity, Bonn, Germany, 22-26 October 2001.

References & Bibliography

Agarwal, Bina, 1992. *The Gender and Environment Debate: Lessons from India*, Feminist Studies 18, No. 1, 1992.

Ahmed, M. Rukunuddin, 1992. *Social Forestry for Environmental Conservation*, "The Bangladesh Observer", Dhaka, 25 July 1992.

Asian Development Bank, 1993. *Forestry Master Plan*, Government of Bangladesh, Ministry of Environment and Forests: 3rd National Forestry Forum Background Paper; published by the Asian Development Bank, Manila, Philippines, March 1993.

Gain, Philip, 1998. *The Last Forests of Bangladesh*, SEHD, Dhaka.

_____ (ed). 2000. *The Chittagong Hill Tracts: Life and Nature at Risk*, SEHD, Dhaka.

Halim, Sadeka. 1998. *Women's Affinity with Nature: Two Perspectives on Ecofeminism* in Dhaka University Studies, Vol. 55, No. 1, June 1998.

_____. 1999.a. *Invisible Again: Women and Social Forestry in Bangladesh*, an unpublished Ph.D. thesis, McGill University, Montreal, Canada.

_____. 1999.b. *Women and Social Forestry in Bangladesh* in "Sociology, Health, Women and Environment", proceedings published by Bangladesh Sociological Association (BSA), Dhaka, October 1999.

_____. 2001.a. *Empowerment of Women: A Way Forward*, paper presented in a conference organized by Bangladesh Social and Economic Forum, May 2001 (to be published by the Bangladesh Unnayan Parishad, Dhaka).

_____. 2001.b. *Role of NGOs and Women's Empowerment: A Case from Bangladesh*, Paper presented at the Second ISTR Asia and Pacific Regional Conference, Osaka, Japan, 26-28 October 2001.

Ishaq, Muhammed. 1975. *Bangladesh District Gazetteers: Chittagong Hill Tracts*, Ministry of Cabinet Affairs, Dhaka.

Khisa, Amarendra Lal. 1963. *Shifting Cultivation in the Chittagong Hill Tracts* (unpublished Master's Thesis), Department of Geography, University of Dhaka.

Roy, Raja Devasish. 1994. *Land Rights of the Indigenous Peoples of the Chittagong Hill Tracts* in Shamsul Huda (ed.), "Land: A Journal of the Practitioners, Development and Research Activists", Vol. 1, No. 1, pp. 11-25, Dhaka, February 1994.

_____. 1996. *Forests, Forestry and Indigenous People in the Chittagong Hill Tracts Bangladesh*, paper presented in a workshop on "The Rights of Tribal and Indigenous Peoples", jointly organized by the Commonwealth Human Rights Initiative and the Minority Rights Group, London, Indian International Centre, New Delhi, February 23-25, 1996.

_____. 1997.a. *Jum (Swidden) Cultivation in the Chittagong Hill Tracts, Bangladesh*, IWGIA, "Indigenous Affairs", No. 1, Copenhagen, 1997.

_____. 1997.b. *The Population Transfer Programme of 1980s and the Land Rights of the Indigenous Peoples of the Chittagong Hill Tracts*, Subir Bhaumik et al (eds.), "Living on the Edge: Essays on the Chittagong Hill Tracts", South Asia Forum for Human Rights, Kathmandu.

_____. 1998. *Land Rights, Land Use and Indigenous Peoples in the Chittagong Hill Tracts* in Philip Gain (ed.), "Bangladesh: Land, Forest and Forest People, SEHD, Dhaka, 2nd ed, Dhaka.

_____. 2000.a. *Salient Features of the Chittagong Hill Tracts Accord of 1997* in Victoria Tauli Corpuz et al (eds.), "The Chittagong Hill Tracts: The Road to a Lasting Peace", Tebtebba Foundation, Baguio City.

_____. 2000.b. *Administration*, in Philip Gain (ed.), "The Chittagong Hill Tracts: Life and Nature at Risk", SEHD, Dhaka.

_____. 2000.c. *Occupations and Economy in Transition: A Case Study of the Chittagong Hill Tracts*, in "Traditional Occupations of Indigenous and Tribal Peoples", ILO, Geneva.

_____. 2000.d. *Land Laws and Customary Rights in the Chittagong Hill Tracts*, paper presented at a workshop on "Land Rights of Indigenous Peoples in Bangladesh", organised by SAFHR in association with RDC, SEHD and Taungya at Dhaka on 28 December 2000.

_____. 2000.e. *The Land Question and the Chittagong Hill Tracts Accord* in Victoria Tauli Corpuz et al (eds.), "The Chittagong Hill Tracts: The Road to a Lasting Peace, Tebtebba Foundation, Baguio City.

_____ & Philip Gain, 1999. *Indigenous Peoples and Forests in Bangladesh*, Minority Rights Group International (ed.), "Forests and Indigenous Peoples of Asia", Report No. 98/4, London, pp. 21-23.

_____ & Sadeka Halim, 2001.a. *A Critique to the Forest (Amendment) Act of 2000 and the (draft) Social Forestry Rules of 2000* in Philip Gain (ed.), " The Forest (Amendment) Act, 2000 and the (draft) Social Forestry Rules, 2000: A Critique", SEHD, Dhaka.

_____ & Sadeka Halim, 2006. "Rights-Based Approaches in Indigenous Forestry in the Chittagong Hill Tracts, Bangladesh: The Village Common Forest Project of Taungya", published by the Lessons Learned Project (LLP) UNDP/RIPP, Bangkok, March 2006.

Roy, Rajkumari Chandra. 2000. *Land Rights of the Indigenous Peoples of the Chittagong Hill Tracts, Bangladesh*, International Work Group for Indigenous Affairs (IWGIA), Document No. 99, Copenhagen.

Sattar, M.A. 1995. *Jhumias Settlement Schemes of Local Forest Department: With Major Focus on Proper Village Land Use Planning and Implementation*; Paper presented at the 'National Workshop on Development Experiences and Prospects in the Chittagong Hill Tracts', organised by ICIMOD et al at Rangamati on 23-25 January 1995.

Scott, James, C. *Seeing Like a State: How Certain Schemes to Improve the Human Condition Have Failed.* New Haven: Yale University Press, 1998.

Shiva, Vandana. *Staying Alive: Women, Ecology and Development in India*, Zed Books, London, 1989.

Sopher, David E. 1963. *Population Dislocation in the Chittagong Hills* in The Geographical Review, Vol. LIII, New York.

Wangza, Devasish Roy, 2010. "Resisting Onslaught on Forest Commons in Post-Accord CHT", in Naeem Mohaiemen (ed), *Between Ashes and Hope: Chittagong Hill Tracts in the Blind Spot of Bangladeshi Nationalism*, Drishtipat Writers' Collective, Dhaka, 2010.

Webb, W.E. & Roberts, R. 1976. *Reconnaissance Mission to the Chittagong Hill Tracts, Bangladesh: Report on Forestry Sector, Vol. 2,* Asian Development Bank, May 1976.

VCF: An Ancient form of Wisdom

Partha Shankar Saha

"Do you see this plant right here? It is called *Dellutti*. It has water in its trunk, which can be used to treat any sort of eye-pain," says Sneho Kumar Chakma, cutting a branch of a leafless plant with his *dao* (hewing knife). Slicing off the ends of its deep green branch he blows into one end. Water gushes out of the other.

As we walk through the forest-covered hills, Sneho Kumar introduces us to many trees, plants, and shrubs. Their uses are as varied as their names. There is the *Jaarbohe* or jungle papaya, which is an effective medicine for a sore throat; *Dubameleni*, whose leaves grounded into a paste, can be used to treat the pain of swollen fingers. *Dubannu* is a plant used to treat the disabled. Tearing some leaves from a plant named, *Sadaraicheya* he says, "The leaves of this plant must be cut and applied to infections, scabs, or irritations. One can bathe in water mixed with this plant and it will cure one from such problems."

We, a group of some ten persons, make our way through a hill forest of Longodu, an upazila in Rangamati Hill District. Sneho Kumar Chakma is our guide. He is 67 years old and lives in Mohajonpur village in Sadar union of Longodu. He is a *jumia*, though he also has some land in the plains. He is well known as a *boidaya* (village doctor or herbal practioner) in his area. He walks barefoot through the jungle, wearing an old *lungi*, a cheap-thin coat, while a hat covers his ears.

It might not be the dark jungles of Africa, but the sun still does sometimes disappear behind the forest canopy. We walk on a rough hilly path. All around us are many types of trees, shrubs, and the bamboos. Suddenly we come across a little stream. During this dry months it hardly has any water. But the huge boulders remind us of the greatness of the falls. "This is something special about the village forests. You will find a stream in every forest", says one of our companions Suchitra Karbari. It is a morning near the end of the winter season. We are going through Modhyahhara village forest in Longodu upazila. Recently these forests have been termed as village common forest (VCF) or village community forest.

These special forests can be found in many villages of the Chittagong Hill Tracts (CHT). A VCF is a forest protected through the collective efforts of people of one or more villages. Maintaining it is the duty of the villagers. It does not belong to one person, but to the entire village. Aside from the widespread use of names for it such as village common forest, common reserves or service forests, etc., the VCF is also known by various names in the many different communities living in the hill tracts. For instance, the Chakma call these forests, *Jaar*, the Tripura *Kalitra*, the Marma *Baam* or *Toykhoyan*. And according to Tanchangya and Khyang it is *Risaab* and *Baam* respectively. In the Mru language the village forests are *Kuyabaam*, for the Bawms it is *Khuya Resereve*, in Pangkhua *Khuya Service*, in Lushai *Service*, and in Khumi *Jumio Pui*. In the language of the Chaks living in Naikhongchhai in Bandarbaan the VCF is *Thing Ding Aka Ara*.

The people of the CHT have protected these forests for hundreds of years. Sneho Kumra Chakma, one of the organizers of Modhyahhara VCF, is a successor of this people. The Modhyahhara VCF is almost 300 acres in area. The people of eight villages—Boradom, Baithapara, Mohajonpara, Bhuiyachhara, Dojorpara, Modhyachhara, Baamechara, and Manikyapara—have worked together to protect this VCF.

The VCFs in the CHT are administered in various ways in different parts of the CHT. Some are taken care of by headmen or *Karbaris*, some by educational and religious institutions. Sometimes committees made up of leaders from one or several villages administer them.

A list of fauna in the VCF has been prepared by Sudatta Bikash Tanchangya and ZuamLian Amlai, two leaders of the Movement for Protection of the Forest and Land Rights in the CHT, a CHT-based organization. Sudatta Bikash Tanchangya claims that they—in a survey with the cooperation of the local people—have recorded 171 species of trees, 62 species of birds, six different types of bamboos, and 22 species of animals in Modhyachhara. They carried out this survey with the help of Taungya, a non-government organization in Rangamati.

Taungya has been working to protect the VCFs since 1996. Its operations include work in 39 VCFs under a project called,

'Conservation of Environment and Societal Development'. Mong Hlaa Miyanto, co-ordinator for the project, says: "The VCFs are currently facing many threats due to increase in population and various social and political pressures. So Taungya is working to preserve these forests and raise awareness among the local people." He informs that they have found a hundred VCFs in Rangamati's Barkal upazila alone. In answer to a question on the total number of VCFs in the CHT, he replies, "There is controversy over the exact number. But I think it must be somewhere between 700 and 800."

ZuamLian Amlai says about Taungya's activities: "The VCFs are located in the unclassed state forest. This is why the hill peoples do not have deeds of ownership for them even though they have been protecting and preserving them for generations. We're working towards official recognition of those who have been truly working to preserve these natural environments."

Dwijen Sharma, noted botanist and nature-lover sees the VCFs as examples of traditional knowledge. "Our forest department should learn from the hill people about how to rightly preserve and protect the forests," says Sharma. "Setting up ecoparks in the name of forest preservation or making gardens full of foreign tree species in the name of social forestry will not protect our forests. This is something the forest department needs to realize."

R. K. Majumdar, Conservator of Forest (CF) at the Rangamati Forest Department, termed the VCF a good initiative. "The Forest Department does not have any institutional connection with the VCFs. But I have seen a few common forests myself," says Mr. Majumder. "The hill people have preserved these types of forest for hundreds of years for their own needs. There are some reasons why NGOs have recently become involved with the VCFs. Our protected forests have been destroyed. I don't hesitate to admit that the forest department is at fault for it. Some vested quarters were behind this failure. And now these parties are blocking any of our current efforts to establish reserved forest parks."

On replicability of the VCFs on a larger scale on the public forest land, Mr. Majumder observed, "The ways in which small VCFs are protected would not be feasible or possible to implement on the large forest lands managed by the forest department."

A typical house in the hills surrounded by remnant forest. © Philip Gain

The Environmental Impacts of Karnaphuli Paper Mill (KPM)

Asfara Ahmed

Paper has been an important instrument in the development of civilization. Its invention was a landmark in human history. It is estimated that the pulp and paper industry directly employs approximately 11,000 people in Bangladesh as well as many more in subsidiary industries. There are 127 registered paper mills in the country (Bangladesh Bureau of Statistics, 2009) mainly concentrated in Chittagong, Rangamati, Khulna, Narayanganj, Gazipur, Sunamganj and Pabna.

Karnaphuli Paper Mill (KPM) is the largest of all paper mills in Bangladesh and it is owned by the state. Approximately 300,000 people are directly or indirectly dependent on KPM for their livelihood (Quader 2003). However, our heavy dependence on paper has come at an enormous price. Our hunger for paper has contaminated our waters, degraded our air, affected our bodies and endangered our forests.

Indigenous varieties of paper such as *afsani* and *tulot* (Fazle 2004) have been produced in Bangladesh for quite some time. However, the domestic paper industry entered a new era of industrialization in 1949, when the Pakistan Government commissioned the construction of KPM. The mill became a symbol of modernization and pride for the entire nation though local communities incurred the social, environmental and economic costs and reaped very little benefit.

Karnaphuli Paper Mill (KPM)

Karnaphuli Paper Mill (KPM) is currently the only operational integrated pulp and paper mill in the country. Its long history of polluting started in 1949 when the construction of the massive industrial unit first began. With a workforce of over three thousand, the industry produces about 30,000 metric tons of paper annually. KPM uses bamboo and tropical hardwood as raw materials to produce 23 different varieties of paper using the bleached sulphate (or Kraft pulping) process.

Brown paper is made using caustic soda and sodium sulfate. Some additives like china clay, dye and binding materials, starch, rosin, and alum are added to the prepared stock of pulp for improving strength and physical properties of the paper.

The waste water contains sulfite liquor, sodium carbonate, sodium sulphide and sodium hydroxide as well as hazardous chemicals such as chromium, sulfur and various acids. KPM sources suggested that the factory is outdated and does not have proper treatment facilities. People along the riverbank have stopped using river water as residents complain of shortness of breath, skin diseases and eye infections. They may also be developing more chronic illnesses such as cancer, which devastate quietly and without warning.

Fish have almost completely disappeared along the stretch of the Karnaphuli River 10 km from the KPM site. The few that are caught reek of pollutants and are not fit for human consumption. The water bears the stench of the 10 tons of industrial waste, which are disposed of in the water every day (Bol 2007). Although a notice for the establishment of an

Waste from KPM flowing through the Karnaphuli river.   © Philip Gain

effluent treatment plant has been served, no action has been taken due to lack of space and funding.

Pulpwood Plantations

Pulpwood plantations have been established to meet the demand for cellulose material for the production of pulp at KPM. The fast growing, high yielding monoculture plantation species are established in place of natural forests, thereby diminishing the biodiversity of indigenous species. The state run mills are the only domestic producers of paper pulp. Large areas of natural forestland are cleared for establishing plantations, land that was once a source of livelihood for many indigenous people. During 2003-2004, the forest production of pulpwood amounted to 670 thousand cubic feet (Bangladesh Bureau of Statistics, 2005).

Hardwoods such as Gamar (*Gmelina arborea*), Shimul (*Bombax ceiba*), Kadam (*Anthocephalus chinensis*), Pitraj (*Amora* species), Koroi (*Albizia* species), etc, are generally harvested for chemical pulping. In ground wood pulping for newsprint, only Gewa (*Excoecaria agallocha*) is used. Various exotic hardwoods such as akashmoni (*Acacia auriculiformis*), mangium (*Acacia mangium*), eucalyptus, etc. are also being grown in Bangladesh (Fazle, 2004) thereby disrupting the ecosystem's delicate equilibrium.

The change in vegetation also has an effect on the water cycle since the rates of water, soil infiltration and evapotranspiration also change (Carrere 1996). Species such as eucalyptus and acacia usually cause hydrological deficits. High yielding industrial production tends to consume quantities of water in direct proportion to rapid growth. The sheer size and magnitude of such plantation projects mean that the problem can affect entire ecosystems with serious impacts on the economy, society, and the environment. These trees are also increasing the ambient air temperature thereby adversely affecting the microclimate of the area (Alam 2004).

In places such as Kaptai Pulpwood Plantation, these species have caused depletion in the ground water level. Many natural springs near Rangamati have dried up because of the plantations. The planting of exotic high yielding species such as acacia and eucalyptus diminish the ground water table thereby depriving the local inhabitants of a source of water for irrigation, sanitation and household consumption. It also inhibits the growth of other species that cannot successfully compete for water, sunlight and nutrients with the exotic species. Species that are associated with the indigenous species also eventually die out.

KPM receives *Gewa* wood from the Sundarbans and large quantities of bamboo from Bangladesh Forest Industries Development Corporation (BFIDC). KPM obtains most of its pulpwood from the Kaptai Pulpwood Division in Rangamati Hill District, which was established in 1978 on 150,000 acres of land and Bandarban Pulpwood division, which was established in 1982 and covers an area of 175,000 acres (Gain 2007). Other sources of pulpwood include leased *M. baccifera* forests, another bamboo forest and an unclassified state forest in the Chittagong Hill Tracts (Saha 1997).

International financial institutions fund many of these pulpwood plantations. The World Bank funded the Forest Resource Management Project for $US49.6 million. The Industrial Plantation component of this project received $US16 million for the production of short rotation saw logs and poles, peeler logs, fuel wood and pulpwood (Gain 2002).

The establishment of pulpwood plantations has many negative environmental consequences. Monoculture brings about the propagation of pests and diseases. Fertilizers, herbicides and pesticides carried by the wind or water contaminate the ecosystem and affect organisms other than the targeted species. Commercial pulpwood plantations also suffer from an imbalanced nutrient cycle due to changes in the recycling of nutrients and in the physical and chemical reactions occurring in the soil. It causes an imbalance between nutrient uptake by plants and nutrient released back into the system by decomposition. Trees such as eucalyptus and pine tend to reduce the action of decomposing agents such as fungi and bacteria. The soil becomes increasingly inhospitable to decomposers due to increased acidification and the introduction of many new chemicals into the ecosystem. Many of the components of hardwood such as lignin, tannin, and waxes are difficult for organisms to decompose. Therefore, the time taken for organic matter to decompose and release nutrients back into the soil also increases (Carrere 1996).

These forests are home to many of Bangladesh's indigenous communities. The pulpwood plantations have eroded the lifestyle, culture, knowledge, history and traditional values of many forest dependent communities. The Khyangs are an ethnic community with a population of about two thousand inhabiting Rangamati's Rajasthali Upazila and their way of life is now threatened due to the encroachment of their ancestral land for a pulpwood plantation expansion program to supply Acacia, Gamar, and Kadam pulpwood to KPM. Once an area is declared a plantation, the customary rights of the local community are nullified. They can no longer rely on the forest's bounty for their sustenance. This land is no longer available for *jum* cultivation thereby forcing the community to flee their homes. There are also cases where the pulpwood division authorities have forcibly planted pulpwood on peoples land (Gain 2002). Thus, the Khyangs of Rangamati have lost the rights to their own land as well as control over their own destiny.

Reference

Alam, M.F. & Mong N. 18 June 2004. Indigenous people in CHT face worst water crisis. The Daily Star [on-line]. Available: http://www.thedailystar. net/2004/06/18/d406181801103.html [June 27, 2007].

Bangladesh Bank Statistics Department. Annual Export Payments 2005-2006.

Bangladesh Bank Statistics Department. Annual Import Payments 2005-2006.

Bangladesh Bureau of Statistics. Handbook on Environmental Statistics, 2005.

Bol, P. One paper mill endangers the lives of 40 thousand people. *Prothom Alo,* March 4, 2007.

Bol, P. Environmental disaster for 10 kilometers around the Karnaphuli. *Prothom Alo,* February 16, 2007.

Carrere, R . & Lohmann L. 1996. Pulping the South. Zed Books.

Fazle, R. 2004. Paper. Banglapedia (CD).

Gain, P. 2007. Stolen Forests. Dhaka: Society of Environment and Human Development.

Gain, P. (2002). The Last Forests of Bangladesh (2nd ed.) Dhaka: Society of Environmental and Human Development.

Quader, S.R. Khulna Newsprint Mill Must Be Saved. The Daily Ittefaq, August 29, 2002.

Quader, S.R. Public Sector Paper Mills: can we not revive them? *The Daily Star,* January 7, 2003.

Saha N., Kawata I., & Furukawa Y. 1997. Alternative Fiber Resources for Pulp and Paper Industry of Bangladesh: Why and What? Journal of Forest Research, 165-170.

Bamboo Resources

Sudibya Kanti Khisa

The Context

Bamboo is an essential component of forest ecosystem in the Chittagong Hill Tracts (CHT). The traditional living and lifestyle of CHT communities, to a large extent, is dependent on bamboo for its variety of uses. It plays a very important role in everyday rural life from cradle to death. It also plays a significant role in the rural economy. Bamboo and rattan serve about 86% needs of household construction materials of the rural communities. It is also linked with other agricultural production systems like betel-leaf production, vegetable cultivation, etc. It indeed, meets the need of IPs for their food supply from the harvest of bamboo shoots during the rainy season when they usually face shortage of food grains particularly in the remote areas. They also sell bamboo shoots and culms in local markets and earn cash. Bamboo is an important raw material for the Karnaphuli Paper Mill (KPM) at Chandraghona. It reduces pressure on the timber consumption, protects natural forests, contributes to alleviating poverty, generates employment and income, improves the environment, and contributes to socio-economic development.

The indigenous peoples (IPs) in rural areas cannot think of a life without bamboo. About a century and half back, Captain Lewin, then Deputy Commissioner of CHT gave a vivid description of bamboo to illustrate the importance of bamboo in the lives of the IPs:

"The bamboo is literally the stuff of life. He builds his house of bamboo; he fertilizes his fields with its ashes; of its stem he makes vessels in which to carry water; with two bits of bamboo he can produce fire; its young and succulent shoots provide a dainty dinner dish; and he weaves his sleeping mat of fine slips thereof. The instruments with which his women weave their cotton are of bamboo. He makes drinking-cups of it, and his head at night rests on a bamboo pillow; his forts are built of it; he catches fish, makes baskets and stools, and thatches his house with the help of the bamboo. He smokes from a pipe of bamboo; and from bamboo ashes he obtains potash. Finally, his funeral pyre is lighted with bamboo. The hill man would die without the bamboo, and the thing he finds hardest of credence is, that in other countries the bamboo does not grow, and that men are in ignorance of it." (Lewin 1869: 28-29).

Bamboo Types and Occurrences

Different species of bamboos occur in CHT and other parts of Bangladesh (Figure-1) both naturally in forests and through cultivation in villages. There are nine genera and more than 33 species of bamboo in Bangladesh, out of which seven occur naturally in the forest areas and about 28 species have been cultivated in the plain land (Banik 2000). Among them *mulibans* (*Melocanna baccifera*) is the most common. Other species— *mitingabans* (*Bambusa tulda*), *orahbans* (*Dendrocalamus longispathus*), *dalubans* (*Neohouzeaua dullooa* Syn. *Schizostachyum dullooa*) and *kalibans* (*Oxytenanthera nigrociliata*) occur sporadically either in association with *mulibans* or in isolation forming small patches of pure strands. The other two species—*latabans* (*Melocalamus compactiflorus*) and *pechabans* (*Dendrocalamus hamiltonii*) are localized only in limited forest areas. A list of the common bamboo species is provided in Table-1. *Mulibans/ egujyabans/paiabans* (*Melocanna baccifera)* is the predominant and major bamboo species growing naturally in different forest areas and constitutes 70% to 90% of the total forest bamboo (Banik, 1998a). It forms pure as well as scattered bamboo vegetation. It covers more than 91,058 ha of natural bamboo forest (Chowdhury 1984, Banik 2000) out of total bamboo area of 287,338. Table-2 depicts the bamboo forest areas of the CHT. *Lotabans/ lodibans (M.compactiflorus)* is a clambering bamboo stretching on the canopy of the tall trees. Apart from naturally occurring forest bamboos, rural communities also plant bamboo in their homesteads and common among these are: *Bambusa vulgaris (Bajyabans/Jaibans/Barialabans), Bambusa polymorpha (Paruabans), Dendrocalamus giganteus (Bhudumbans)* in the CHT and *B.balcoa (Baruabans), B.longispiculata (Tallabans/Makiabans), B.nutans (Malbans), B.arundancea (Katabans) and D. strictus (Lathibans)* in other parts of the country. Some forest bamboo species like *Dendrocalamus longispathus (Orahbans), M.baccifera* and *B. tulda* are also cultivated in homesteads. *Dendrocalamus hamiltoni (Pechabans), Melocalamus compactiflorus (Latabans) and Schizostachyum dullooa (Dolubans)* are the threatened species in Bangladesh (*Banik 1998a)*. However, community-based conservation of *S. dullooa, D. longispathus and M. baccifera* is seen in many areas of the CHT. *Dendrocalamus giganteus (Bhudumbans) and Bambusa polymorpha (Paruabans)* are cultivated in Buddhist temples in the CHT, Chittagong, and Cox's Bazar.

Bamboo uses, production areas, estimated demand and supply

Bamboo has a versatile use (Table-!). It is used for rural house construction, scaffolding, ladders, mats, baskets, fencing, tool-handles, pipes, toys, fishing rods, fishing traps, handicrafts, utensils, furniture etc. and several other articles of everyday use (GoB 1993). In some parts of the country, the bamboo leaves are used as thatching materials. It is one of the raw materials for the pulp and paper industry. It is also planted for hedges and landscaping. Bamboo groves also act as a wind break and prevent soil erosion. The young tender shoots of bamboos are delicious vegetables. These young shoots locally known as *banskorol* are much eaten by the IPs during the rainy season. Considering the wide ranging use as construction material, it is called the "poor man's timber". As a matter of fact, life in the rural areas cannot be conceived of without the use of

A house in a Mongu Para, Khumi village in Roangchhari upazila (Bandarban), constructed entirely of bamboo.   © Philip Gain

bamboo in the present stage of the economy and for many years to come. A total of 706.3 million bamboo culms were used for making these items in Bangladesh in 1993 (Banik 2000). Dry bamboo leaves are extensively used as fuelwood and green leaves as fodder in the rural areas. Bamboo rhizomes are sold to brick kilns at the rate of 600 to 800 Taka (USD 7 to 10) per ton (Banik, 2000). The young shoot of *Melocanna baccifera, Dendrocalamus longispathus, D. hamiltonii and Bambusa tulda* are used as vegetable. The yield of edible-shoot ranges from 3 to 30 tons per hectare per year (Banik 2000). However, no information is available about total

Bamboo growing areas in CHT and other parts of Bangladesh

production. A valuable medicine, Tabashir or Banslochan is occasionally found inside the culm internodes of *Bambusa spp, Dendroclamus strictus and Melocanna baccifera*. It is also an important ingredient for preparing famous Ayurvedic medicine, "Chawanprash" commonly used as a cooling energy tonic, aphrodisiac, cure for chronic cough and old age weakness (Banik 2000). *Bambusa polymorpha and B.vulgaris variety striata* are used as ornamental bamboo.

In Bangladesh, 20% of the bamboo supplies are from forest and 80% from the village source. By 2013, the estimated bamboo supply from natural forest is 49.4 million culms whereas from villages it is 527.5 million (Banik 1998a) against the estimated demand of 901.5 million culms (domestic demand 730.1 million, urban housing 47.0 and industrial demand of 124.4 million culm) with the shortfall of 324.6 million culms. In Karnafuli Paper Mill alone 28.3 million culms (45,000 ADT x 625) were annually used before its BMRE (Balancing, Modernization, Rehabilitation, and Expansion). After BMRE, production has increased to about 55,000 ADT annually. Therefore, 34.4 million culms (55,000 ADT x 625) are in use in KPM from 1994-95 onwards.

Table 1: Some common bamboos of the CHT and their uses (Alam 1992; Banik 1998a)

Scientific name (local name/s)	Uses
B. polymorpha Munro. (Paruabans, Wapiabans)	Construction of houses, agricultural implements, pulp and paper, fiber board, young shoots are edible, boat-plying rods, ornaments, fishing implements, floats, pans, traps, stakes for plantations, cart shed roof, containers for storing grain, horticultural pursuit, chicks for doors and windows, fencing, tool handles, fodder, fuel, cooking utensils, basket making, thatching and roofing etc.
B. tulda Roxb. (Mitingabans, Aille, Keyitta, Tallabans, Taralabans)	Construction of houses, agricultural implements, pulp and paper, fiber board, young shoots are edible, boat-plying rods,

ornaments, fishing implements, floats, pans, traps, stakes for plantations, cart shed roof, containers for storing grain, horticultural pursuit, chicks for doors and windows, fencing, tool handles, fodder, fuel, cooking utensils, basket making, thatching and roofing etc.

B. tulda Roxb.
(Mitingabans, Aille,
Keyitta,
Tallabans, Taralabans)

Wall construction and rural housing, thatching and roofing, making toys, mats, screen, wall plates, hats, basket, pulp and paper, boat roofs, stakes for plantations, containers for storing grain, haystack stabilizer, shoots for food, *hooka* pipes, fencing, tool handles, fodder, fishing rods, fuel, furniture, water and milk vessels (*chunga*), cooking utensils, basket making etc.

B. vulgaris Schrad ex
Wendi.
(baijjyabans/ barialabans)

Construction works, jaundice treatment, rickshaw hoods, handicrafts, bridges, boat-mast, cordage, ornaments, fishing implements, floats, pans, traps, containers to administer medicine to animals stakes for plantations, cart shed roofs, containers for storing grain, ladders, protection during grain pounding, cart yokes, scaffolding, cradles, cremation, coffins, haystack stabilizer, trays for silk worms, tent poles, chicks for doors and windows, country tiles, *hooka* pipes, fencing, fodder, fuel, hedges, agricultural implements, cooking utensils, basket making, props, scaffoldings, bridge

	making, boat-masts, cottage industries for making toys and handicrafts, pulp and paper. Young shoot is edible.
Dendrocalamus giganteus Munro *(Budhumbans, Kanchanbans, Rajabans)*	Construction, hedges, water and milk vessels (*chunga*), basket making etc.
Dendrocalamus longispathus (Kurz) Kurz *(Khang, Orahbans, Rupaibans)*	Construction, musical instruments chicks for doors and windows, floats for timber rafts, fodder, fuel, basket making, thatching and roofing, furniture and food grain containers, raw materials for pulp and paper. Young shoots are edible.
Gigantochloa andamanica Kurz Synonym: *Oxytenanthera nigriciliata* Munro *Oxytenanthera auriculata*(Kurz) Prain *(Kalibans, kaliseribans, klijori)*	Construction, shoots for food, leaves as fodder, fuel, hedges, walking stick, basket making, etc.
Melocalamus compactiflorus(Kurz) Benth. *(Latabans/Dharalabans)*	Not known
Melocanna baccifera (Roxb.) Kurz Synonym: *Melocanna babmusoides* Trin *(Mulibans, egujyabans, paiyyabans, bajalibans, nail, tarai)*	Construction works, young shoots for food, pulp and paper, cottage industries, umbrella handles, boat plying rods, fishing implements, pans, traps, boat roofs, stakes for plantations, cart shed roofs, containers for storing grain, horticultural pursuit, chicks for doors and windows, fencing, tool handles, floats for timber rafts, fodder, fuel, seed for food, furniture, water and milk vessels (*chunga*), mats, cooking utensils, basket making, thatching and roofing, walling and rural housing etc.

Schizostachyum dullooa Gabble) Majumder *syn.Neohouzeaua Dullooa (Gamble) Camus (Dolubans)*	Construction works, general utility, umbrella handles, fishing implements, pans, traps, pan trays, stakes for plantations, cart shed roofs, containers for storing grain, pipes, chicks for doors and windows, fencing, floats for timber rafts, fodder, fuel, cooking utensils, loading vessels, thatching and roofing, walling and rural housing, making mats, baskets, and novelty items, etc.
Dendrocalamus hamiltonii (Pechabans)	General utility, construction, shoots for food, fodder, fuel, mats, cooking utensils, etc.
Bambusa longispiculata Gamble ex. Brandis (Tallabans/Makiabans)	Housing, roofing, making toys, mats, screen, wall plates, hats, basket, containers, pulp and paper, etc.

Table-2: Bamboo forest areas of CHT (Anon 1963; De Milde et al. 1985; GoB 1993)

Bamboo forest areas	Area (ha) and estimated year	Remarks
Kassalong Reserved and Reingkhyong Reserved Forests	2,41,631(1961- 63), 74,084 (1985), 71,042 (1993)	No record latest inventory was carried
Sangu and Matamuhari Reserved Forests.	31,260(1961), 24,606 (1985)	out
Banbarban USF	14791 (1961), 14791 (1985)	

Bamboo Biology

Bamboo belongs to the family *Poaceae (Gramineae)* of plant taxa. It has tree like habit and can be characterized as having woody, usually hollow culms, complex rhizome and branch systems. There are an estimated 1,000 or so species of bamboo belonging to about 80 genera in the world. Of these about 200 species belonging to approximately 20 genera are found in

South East Asia. So far, 28 species and one variety of bamboo under seven genera have been recorded from Bangladesh (Alam 2001, Gamble 1896, Hooker 1897, Prain 1903, Brandis 1899, and Bor 1940). These genera are: *Bambusa, Dendrocalamus, Gigantochloa, Schizostachyum, Melocalamus, Melocanna,* and *Thyrostachys.* Most of the species belongs to the genera *Bambusa* and Dendrocalamus. The remaining genera are reported to have only 1-2 species each. Among 28 species of bamboo, seven are found to grow naturally in the forests of Bangladesh. Forest bamboo species are generally thin-crusted (thickness <1 cm) in nature and shorter than village grove bamboo (4-15 m long). The remaining species referred to as village bamboo have been cultivated in the plains land. At present, 80% of the total bamboo supply comes from village bamboos. Village bamboos are generally thick-walled (thickness 1.5-2.5 cm) in nature and taller than the forest bamboos (17-35 m long). Bamboos have a very long life (from seed to seed), are very fast growing woody species and are harvestable within 3 to 5 years after planting. A bamboo culm (aerial stem) grows annually from the underground rhizome system. A bamboo culm goes on producing new culms every year, and matured culms can be harvested perpetually every year with minimum management efforts. Most of the bamboo species in Bangladesh are 'Sympodial' bamboo. But the most prominent bamboo in natural forests, *Melocanna baccifera* (Mulibans), arising singly in a variety of distances from a common creeping rhizome is not true of 'monopodial' categories.

Bamboo Ecology

Bamboos have an extremely wide range of global distribution from lowland to 4,000 m altitude that occurs in the tropical, sub-tropical and temperate regions of all continents except in Europe and Western Asia. The distribution and growth of bamboos are influenced by temperature. High temperature usually acts favorably on the growth of bamboos. The mean annual rainfall in bamboo growing areas is over 1,000 mm. Bamboos cannot stand water logging. Light is an important factor in respect to the distribution of natural bamboos. Bamboos grow in forests in varied conditions. They grow as pure and as under-storey in timber areas. Most bamboos are found in well-drained, sandy loam to clay loam soil with a pH range of 5.6 to 6.5. In the CHT there are two distinct bamboo production areas: the forests and villages. All natural forest bamboos in the CHT (Table-2) other than *Mulibans* grow as an under-storey of moist evergreen and semi-evergreen forests. But *Mulibans* grows as pure brakes as its rhizome type root system is a factor in spreading in clear areas. In villages, bamboos are cultivated in small clusters in and around the homesteads or as small groves. The distribution of village bamboos depends on the human factors such as utility, availability of propagating materials locally, and human migration. In village groves, usually thick-walled bamboos like *Bajyabans, Mitingabans* and even thin-walled *Mulibans* are planted and cultivated for domestic consumption and as raw materials for cottage industries.

Bamboo Phenology

The bamboo clumps produce culms generally from the month of May or June and continue for 6 to 7 months ending in October or November (Banik 1993b). The shoots from rhizome develop during the pre-monsoon, and grow during the rainy season. The Elongation of the culms takes place during the rainy season and continues till the post rainy season. The species exhibited different durations of shoot emergence periods varying from 4 to 8 months. The total culm elongation periods of *Bhajyabans (B. vulgaris)* are 75-85 days, and 55-60 days for *Mulibans (M. baccifera)*. The rate of daily culm elongation varies from 40 to 70 cm depending on the species. The natural mortality of emerging culm is 28-69 percent in thick-walled, tall species, compared to lower rates (9-37%) in thin-walled and small size bamboo species. The life of culms in the clumps is long (10-13 years) in *Bhudumbans (D. giganteus)*, while it is short (5-10 years) in most of the thin-walled bamboo species. In the clumps of *Mitingabans (B. tulda)* and *Bhajyabans (B. vulgaris)*, production and growth (height and diameter) of full grown culms gradually increase up to the fifth year of planting then stabilize and gradually decline. In most of the species, clump expansion takes place up to 8-10 years whereas in *Mulibans (M. baccifera)* the clump continues to expand even after the age of 15 years. Table-3 shows the clump and seed characters of some bamboo species.

Bamboo Flowering

Flowering is a very interesting aspect of the life of bamboos. Most bamboo species flower at a stage of their life and die after flowering. Some bamboos flower every year, others at short intervals, and the majority at long intervals. The entire population of a species that blooms after long intervals flowers simultaneously. Such a peculiar behavior of bamboos often creates ecological, economic, and social problems. The possible next flowering year of the major bamboo species of the country can be forecasted on the basis of their estimated inter-seeding (flowering cycle) period. Bhajya/bariala (*B. vulgaris*), the most common bamboo species cultivated in the villages, usually flower very sporadically or in isolated clumps after many years. But they do not produce any seeds. The estimated ranges of flowering cycles of different bamboo species (Table-4) of Bangladesh were found to be within 20-80 years but the majority have 30-50 year cycles (Hasan 1973, Banik 1980, 1992). Banik (1998) reported that *Melocanna baccifera* exhibited a flowering cycle of 30+5 and other was 45+5 years. *Bambusa tulda* frequently flowers sporadically and also exhibits gregarious nature of flowering after 20+5 years. *Dendrocalamus longispata* often flowers sporadically and also sometimes a gregariously after 30+2 years. Flowering in *0. nigrociliata* and *N. dullooa* is sporadic and also occasionally gregarious after 47+3 and 45+2 years respectively.

When *Mulibans* flowering starts in an area, a 'flowering wave' continues for about 6 to 10 years. So, all bamboos in an area do not die in a single year but continue for 6-10 years in different areas in different years. There also exists different genotypes of the species in a geographical region that protects the species' extinction. The gregarious flowering of bamboos produces large quantities of seeds, which in turn causes sudden population explosion in rats. However, the quantity of seeds available for rats diminishes soon due to germination of seeds after the rains. The resultant short supply of bamboo seeds on the one hand and a large population of rats on the other, makes rats heading towards *jum* fields and other crop lands in the adjoining areas and causing widespread loss of the crops. Such a chain of events has the potential to cause famine. Mostly three bamboo species--*Mulibans(Melocanna baccifera), Dendrocalamus hamiltoni, and mitingabans (Bambusa tulda)*--are responsible for increase in rodent populations synchronized with their flowering.

During 2007, *Mulibans* gregariously flowered in Sajek, Thanchi, Ruma, Bilaichhari, Barkal, and other areas in the CHT. Its flowering was first recorded during 1863 to 1866. It also flowered sporadically during 1901 to 1905. Thereafter a gregarious flowering took place and continued for two years during 1960 and 1961. During this time the species flowered in the CHT like a wave covering an area of about 1,000 sq. miles in four (1957-1961) years' time (Hasan 1973). Comparing the present flowering with that of the late 1950s and early 1960s it is seen that the flowering cycle of *Mulibans* in the CHT is 50 ± 5 years. *Kalibans* flowered in Pablakhali in 1978; recently it flowered in the CHT and its estimated flowering cycle is 30 years.

Immediate effects of flowering: The flowering results in the danger of fire hazard due to the sudden and huge stockpile of dry bamboos; the sudden explosion of rodent population due to availability of excess food from bamboo seeds; a sudden shortage of seeds due to germination and the rodents' invasion of nearby fields destroying agricultural food crops; food shortage and shortage in supply of bamboo shoots in the affected areas; scarcity of bamboo leading to hardship for the people who are dependent on bamboo for their livelihood; and the fear of epidemic outbreak.

Impacts of bamboo flowering on the lives and livelihoods of the communities in the affected areas: *Mulibans*, when flowering, produces huge quantities of fleshy fruits (seed), which are readily devoured by cattle, elephants, bison, deer, pigs, and other animals. The fruit is edible and used as an additional source of food during food scarcity. Its flowering starts in September-October, immediately after the rainy season. After about two and a half months the floral shoots start blooming during November-December. Peak flowering takes place during December-February when the leafless flowering culms in a clump look like a giant inflorescence. Fruit (seed) setting and maturation take place within next four to five months in April and May. "Seed" production is optimum during May to June and poor from later part of September to November. The fruit-shed attracts the predators, mostly rats. A significant increase in rodent population takes place as they get plenty of food supply from fruits. When seed shortages occur, the food supply for rats decreases, then rodents search for alternative supply of food, often from crops and granaries destroying crops like paddy, papayas, chili, and gourds etc. By the end of summer, there are enough seeds for the increase in rat population. Then the rainy season starts. The sudden decline in food diverts thousands of rats towards cultivated *jum*, and ultimately to the paddy fields in the valleys and plains. This leads to a food crisis and causes not only famine for the time being but also a depletion of bamboo resources at least for four to five years.

Impacts on seed security and agro-biodiversity: The rapid increase in rats, said to be a "rat flood", damages not only crops but also disrupts the seed security of *jumias* who usually maintain their own seed stock from their previous harvests. Following the "rat flood" in the nearby *jum* and paddy fields, farmers cannot stock their seeds for the next sowing. Along with the food crisis this also affects the future *jum* production.

Impacts on the habitat and vegetation: *Muli*bans occurs mostly in pure brakes (an area thickly overgrown with any particular species/thicket—a dense growth). Unlike many other bamboos *Mulibans* dies completely including aerial culms and underground rhizomes. Complete dying results in denudation of habitats. This complete dying of whole bamboo clumps has definite impacts on the habitat. If a land regenerating with bamboo seedlings is cleared and burnt for *jum*, it hampers its normal regeneration process. Otherwise the land will revert to a bamboo forest of mixed vegetation. Accidental fire may cause damage to regenerated bamboo and tree growth. Grazing also causes damage to young bamboo seedlings.

Need for awareness about bamboo flowering: For sustainable management of bamboo forests it is important that people become aware of bamboo flowering, a recurring natural phenomenon. Considering the ecological significance and vast economic potential of bamboos in the region, extension and awareness about bamboo flowering needs to be given renewed thrust by adopting a "bamboo flowering policy" with the following aims and objectives:

■ Protection and preservation of bamboo forests and bamboo re-growth areas for sustained productivity and environmental security for the people.

■ Regeneration of flowering affected areas after completion of the felling operation of flowering bamboo culms and clumps.

■ Integrated effort for rodent control to prevent famine and health hazards. In order to control the increase in the rodent population, the concerned departments need to draw detailed contingency plans for procurement, storage and distribution of rodenticides, and for creating awareness among the people about the need of preventing such rat population increase.

■ Creating adequate storage facility to stock food grains to meet any exigency and thrust be given at the farmer's level to take up rat preventive measures for in-house storage structures/granaries. Necessary inputs should be given to the grain producers for proper storage. Government may chalk out detailed plans for making arrangements for procurement, storage and quick distribution of food items in the event of a famine. They need to have safe storage godowns and necessary infrastructure for timely and efficient distribution of food supply. If the famine does not occur, the plans for utilization of the extra food stock and saving them from rodent damage should also be kept ready.

- Raising awareness among the communities affected by bamboo flowering about this impending problem and issue a set of *"dos and don'ts"* for the communities.

Bamboo policy and legislation: Bamboo and rattan fall under the definition of "forest produce" as per the Forest Act, 1927 and are categorized as "minor forest products" from the utilization point of view although about 90% needs of rural housing, construction, thatching, household articles and firewood of the rural populace are being met by them. The provisions, rules and regulations under the existing Forest Act, 1927 control and regulate access to the bamboo and rattan in the government forestland. Illegal felling or extraction of these produce is also punishable under the said Forest Act, which has an impact on the protection and conservation of these valuable resources. The movement of these produce is also controlled or regulated by the existing transit rules in the country. Existing cutting rules of bamboo are contained in Annexure-"A".

The Way Forward

Bamboo forests are now severely degraded in the CHT due to the continued unsustainable harvest and misuse. The CHT presently suffers a deficit in bamboo supply. The shortfall has increased alarmingly due to large-scale death of forest bamboo due to gregarious flowering. This situation is severely affecting the lives and livelihoods of the CHT communities. However, there exist many opportunities for replenishment of bamboo resources along with biodiversity management and livelihood enhancement. There is also a tremendous benefit of planting bamboo as a plantation crop.

Considering the ecological, productive and economic services that bamboos provide, its full ecological and economic potential needs to be recognized for sustainable development of the CHT region and also for the livelihood security to the CHT communities. To ensure future sustainable supply of this valuable resource, it needs an immediate management plan with strengthening of bamboo action research as a backstops. The following recommendations may be adopted as the way forward:

- Development of bamboo management plan in consultation with stakeholders and promotion of bamboo cultivation either through natural regeneration by allowing the bamboo seeds that drop on the ground to germinate and grow or through aided regeneration, or artificial generation by planting of seeds as it is the cheapest planting material. *Mulbans* flowers gregariously once in 50 years producing huge quantity of seeds. Millions of seeds become available after its flowering. As whole bamboo brakes die after flowering, a vast chunk of this land becomes vacant for planting. This chance is available only once in every 50 years. However, bamboos can be propagated vegetatively.

- Integration of bamboo cultivation with timber and other non-timber forest products. Different bamboo production models like pure bamboo plantation, bamboo+timber tree plantation or bamboo+timber tree+non-timber (medicinal plants) plantations can be designed based on local needs, access to markets. One of the aspects of bamboo ecology is that it grows as an understory with other vegetation. So, bamboo plantations can be made in an admixture of timber and non-timber species as mixed planting always ensures more sustainability. Land particularly the D-type or C-D type (73% of the CHT land area) may be used for bamboo cultivation.

- Development of value added bamboo based commodities may be promoted as bamboo is a raw material for versatile commodities from kitchen ware, agricultural implements, handicrafts to industrial products. The various cottage industries for *agarbati* (fragrance) sticks, toothpicks, bamboo mats etc. require bamboo. Markets need to be explored and developed with development of value added bamboo commodities. Linking of the producers with the consumers through efficient marketing will ensure sustained production of bamboo. Employment opportunities and cash generation by developing commodities from selective species will reduce pressure on other resources and thus will help in conserving biodiversity. Value added bamboo products have very good domestic and international market potentials. Different indigenous communities have their own traditions, cultural brands and craftsmanship that can be utilized in developing value added products with market potentials.

- Utilizing traditional local institutions for bamboo resource management: One of the essential advantages of CHT is the existence of traditional institutions headed by" headmen (head of a mouza) and karbaries (head of para)" among the IPs in the CHT that will help in developing bamboo resources and their management.

- Integrating and incorporating bamboo production plan in the existing village common forests (VCFs). Most of IP communities in some *paras* (villages) and *mouzas* (consisting of few *paras*) still maintain VCFs around the village or in the vicinity that is managed by the IPs. These VCFs are usually maintained for ecological services like maintenance of mini watersheds, protection of the village from heat and fire. VCFs also supply non-timber forest products and also timber for use of the community. Bamboo production can be incorporated with this existing norm of VCF in the essence of community resource development and management. Integration of bamboo as a major component in existing VCF will help easy production and conservation of this resource.

Literature consulted

Alam, M. K. 2001. *Bamboos of Bangladesh: a Field Identification Guide.* Bangladesh Forest Research Institute, Chittagong. 35 pp

-------. 1995. *Melecanna baccifera* in Dransfield, S. and Widjaja, E. A. (eds.) Plant Resources of South-East Asia No. 7. Bamboos. Backhuys Publishers, Leiden.pp.126-129.

-------. 1992. A Note on the Taxonomic Problems, Ecology and Distribution of Bamboos in Bangladesh.J.Amer.Bamboo Soc.9 (1&2): 1-7 pp.

-------. 1982. A Guide to Eighteen Species of Bamboos from Bangladesh. Bull Taxonomy Series. Forest Research Institute, Chittagong. 29 pp

Anon. 1964. Chittagong Hill Tracts Forest Inventory Survey (1961-63). Kassalong and Reinghyong Reserve Forests. vol. 1, Forestal, Canada, Project no F 334.

Banik, R.L. 2000. Silviculture and Field-Guide to Priority Bamboos of Bangladesh and South Asia. Publication of Bangladesh Forest Research Institute, Chittagong. 1998a. Reproductive biology and flowering populations with diversities in *muli* bamboo, *Melocanna baccifera* (Roxb.) Kurz. Bangladesh Journal of Forest Science 27(1): 1-15.

-------. 1998b. Bamboo resources, management and utilization in Bangladesh. Country report presented in the Bamboo Training Course/ Workshop in Kunming and Xishuangbanna, Yunnan, China, 10-17 May, 1998.   http://www2.bioversityinternational.org/publications/Web_ version/572/ch22.htm

-------. 1998c. Conservation and propagation challenges of bamboo and rattan resources in Chittagong Hill Tracts, pp. 103-111. In : Banik, R.L. Alam. M. K.S.J. Pei and A. Rastogi (eds.). Applied Ethnobotany BFRI-UNESCO-ICIMOD. Chittagong, Bangladesh.

-------. 1998d. Bamboo genetic resources of Bangladesh, pp 1-32. In : Vivekanandan, K., A.N. Rao., and Ramanath Rao(eds.). Bamboo and genetic resources in certain Asian countries. IPGRI-APO, Serdang, Malaysia.

-------. 1998c. Management of wild bamboo seedlings for natural regeneration and reforestation, pp 92-95 in Bamboo-current research. Proc. of the III International Bamboo Workshop. Cochin, India, 1988.

-------. 1997. Bamboo resources of Bangladesh, pp 183-207 in Alam, M.K., Abrned, F.U and Amin, S.M.R.(eds.). Agroforestry: Bangladesh perspective APAN-NAWG-BARC. Dhaka.

-------. 1995. A manual for vegetative propagation of bamboos, 66p. INBARTechnical Report # 6. IDRC/IPGRI/FORTIP. 1995.

-------. 1994a. Distribution and ecological status of bamboo forests of Banglades. Bang. Jour. of For. Sci. 23(l&2), pp 12-19, 1994.

-------. 1994b. Studies on seed germination, seedling growth and nursery management of Melocanna baccifera (Roxb.) Kurz. pp 113-119. In: Proc. 4 Intl. Bamboo Workshop on Bamboo in Asia and the Pacific; Chiangmai, Thailand, Nov. 27-30,1991.

-------. 1993a. Bamboo. Forestry Master Plan of Bangladesh. Pp. 62 (8 Apendix) in Asian Dev. Bank (TA.No. 1355-BAN). UNDP/FAO BGD.

-------. 1993b. Periodicity of culm emergence in different bamboo species of Bangladesh. Annals of Forestry 1(1), pp.13-17

-------. , Islam, S.A. M.N., and Hadiuzzaman, S. 1993. In vitro regeneration of multiple shoots in three bamboo species. Plant Tissue Culture 3(2), 101-106 pp

-------. 1992. Bamboo. Forestry Master Plan of Bangladesh. Asian Dev. Bank (TA No.l355-BAN). UNDP/FAO BGD 88/025. 62 pp (8 Apendix).

-------. 1991 Trial on the tissue culture ofMelocanna baccifera (Roxb.) Kurz. p 135. Proc. of IV Intl. Bamboo Workshop, IDRC, Chiangmai, Thailand (Abstract).

-------. 1989. Recent flowering of muli bamboo (Melocanna baccifera) in Bangladesh: An alarming situation for bamboo resource. Bano Biggyan

Patrika 18(1&2):65-68

-------. 1988. Management of wild bamboo seedlings for natural regeneration and reforestation. Pp. 92-95 in Bamboo-Current Research. KFRI, Peechi, India. IDRC, Singapore

-------. 1987a. Seed germination of some bamboo species. Indian Forester 113(8), 578-586 pp.

-------. 1987b. Techniques of bamboo propagation with special reference to prerooted and prerhizomed branch cuttings and tissue culture. pp.160-169 pp in Rao, A.N., G. Dhanarajan and C.B.Sastry(eds.).1987. Recent Research on Bamboos. Proceedings of the International Bamboo Workshop held in Hangzhou, Peoples Republic of China during October 6-14, 1985.

-------. 1984. Macro-propagation of bamboos by pre-rooted and pre-rhizomed branch cuttings. Bano Biggyan Patrika 13(l&2), 67-73 pp.

-------. 1980. Propagation of bamboos by clonal methods and by seeds. 139-150 pp. *In*: G. Lesserd and A. Chouinard, (eds.). Bamboo Research in Asia. IDRC, Ottawa, Canada.

Boa, E. R. and Rahman, M. A. 1987. Bamboo blight and bamboos of Bangladesh. *Forest Pathology Series, Bulletin 1* Bangladesh Forest Research Institute, Chittagong. pp 43.

Bor, N.L. 1940. Flora of Assam.Vol.4, 480 pp (reprinted 1982)

Brandis, D. 1899. Biological notes on Indian Bamboos. Indian Forester, 25:1-25.

Chakraborty, Nalini Kanta. 2005. BANS SAMPAD (Bamboo Resources) in Bengali. Agartala, India.

Choudhury , M. R. 1984. A study on supply and demand of bamboos and canes in Bangladesh. FAO-UNDP Project BGD/78/010. Field Document No. 9, Dhaka.

De Milde, R; Shaheduzzaman, M.; and Chowdhury, J. A. 1985. The Kassalong and Reingkhyong Reserve Forests in the Chittagong Hill Tracts. Assistance to the Forestry Sector, Bangladesh, FAO/ UNDP Project, BGD /79/017. Field Document No. 10.

Gamble, J.S. 1896. The Bambuseae of British India. Annals of Royal Botanic Garden, Calcutta, 7:133 pp.

GoB. 1993. Forestry Master Plan (FMP), Statistical Data. ADB TA No. 1355-BN. UNDP/FAO BGD/88/025.

Hasan, S.M. 1973. Seeding behaviour of Bangladesh bamboos. Bano Biggyan Patrika 5(2): 21-36.

HKI. 2008. Recommended Responses to the Rodent Crisis in the Chittagong Hill Tracts :Evidence from Food and Nutrition Survey. Prepared for UNDP, Bangladesh by Helen Keller-Bangladesh. 27 pp.

-------. 2008. Needs Assessment Report on Bamboo Flowering, Rat Infestation and Food Scarcity in the Chittagong Hills Tracts, Bangladesh prepared for USAID.33 pp.

Hooker, J.D. 1987. Flora of British India.Vol.7.L. Reeve and Co.Ltd. Kent, p.842.

Lewin, T.H. 1869. The Hill Tracts of Chittagong and the Dwellers therein with Comparative Vocabularies of the Hill Dialects, Calcutta, Bengal Printing Co. ltd.

Lewin, T.H. 1856. *Wild Races of the Eastern Frontier of India.* Published by K.M. Mittal, Mittal Pulications, India.

MSF. 2008. Food Security Assessment Report-Chittagong Hill Tracts, Sajek Union. Mendicins Sans Frontiers(MSF)-Holland, Bangladesh.

Nuruzzaman, Md . National Report on the State of Bamboo and Rattan Development in Bangladesh. http://www.inbar.int/documents/country%20report/Bangladesh.htm

Prain, D. 1903. Bengal Plants. 2:663-1319. Calcutta

UNDP, 2008. Scientific Assessment Report on Bamboo Flowering, Rodent Outbreaks and Food Security: Rodent ecology, pest management and socio-economic impacts in the Chittagong Hill Tracts, Bangladesh.50 pp.

Zashimuddin, Mohammad. 2005. Bangladesh Country Report on Bamboo Resources. Beijing, 9 May 2005. Global Forest Resources Assessment-2005. FAO Working paper-112. INBAR (International Network for Bamboo and Rattan).

Bamboo Flowering, Rat Flood and the Grief of the *Jumias*

Partha Shankar Saha

Bamboo flowering, a death-knell for bamboo groves, in
Kawkhali.   © Partha Sankar Saha

Aung Khei Marma (70), father of six children, lives with his wife in Majher Para, a hill village of Kowkhali Thana in Rangamati Hill District. All his children are grown up now and have their own families. The harvest from his *jum* plot feeds him and his wife just for half of the year. Aung Khei and his wife collect and sell bamboos from the forest for an income to feed themselves for the other half of the year.

Bamboo or *owa* is a symbol of life to him and other hill people. Aung, a *jumia* (slash-and-burn cultivator) says, "Owa Asaku *Kichiay*" or Bamboo is the staff of my life.

However, bamboo seemed to have been at the center of a catastrophe for Aung Khei Marma and many others in his hill village of some 150 families (all Marmas) since June 2007. It began with the blooming of their bamboo groves. Once a bamboo grove begins to flower and bear fruits, it is bound to die soon. Experience shows that bamboo flowering is followed by massive attack of rats, popularly known as rat flood. The attack of rats is a nightmare for the *jumias*. Aung Khei sowed 24 kilograms of paddy seeds in 2007. The rats fed on three-fourths of the harvest. Others of his village suffered the same fate. They all now await staple to be supplied in aid for mere survival.

A bamboo fruit (little larger than a golf ball) that follows the flowering contains a high level of protein and attract the rats. Feeding on bamboo fruits the rats gain unusual reproductive ability. Consequently, the rat population increases rapidly leading to what is known as 'rat flood'. The rats in massive numbers not only feed on bamboo fruits, they quickly consume other food available in the jungle. *Jum* with pumpkins, potato, paddy, and other crops become particular attraction to these massive numbers of rats.

Some of several hundred species of bamboo flower every year, and some irregularly, causing no harm. Yet, few species flower after a long interval and continue blooming for two-three years.

Mulibash (*Melucanna baccifera*) that constitutes 70% of bamboo stocks in the CHT began to bloom in 2006. This heralded a fear especially for those who depend on *jum*. Flowering of *mulibash* takes place every 50 years. This time this particular species of bamboo in Mizoram neighboring the CHT began to flower in 2005. Sajek in the CHT was one of the first areas to witness the bamboos blooming.

The United Nations Development Program (UNDP) estimated that some 130,000 people would be in tremendous food crisis in the CHT in 2008 due to loss of crops.

The hill people see the blooming of bamboos as a sign for famine. However, the authorities were hardly prepared to face the food crisis the common man feared. "Actually, we had no preparation to face this. At least one year back, I informed the higher authorities of the Department of Agricultural Extension (DAE) in Rangamati and Dhaka about the danger that comes with the flowering of bamboos. But they did not pay attention," says Kajal Talukdar, district training officer of DAE.

The DAE of Rangamati estimates that paddy of 2,561 ha of land is lost because of the rat flood. The Forest Department is yet to calculate the extent of bamboos lost because of rat flood.

The Conservator of Forest (CF) for the Rangamati circle RK Majumdar thinks that the blooming of bamboo is a natural phenomenon. He thinks it is good for the regeneration of bamboo. He says, "We are trying to tell people not to collect bamboos from flowered ones. We also sensitized people so that they don't burn the flowering bamboo." However, he admits that the Forest Department did not take any step to tackle the rat flood.

A member of the Chittagong Hill Tracts Regional Council, Rupayan Dewan, blamed the ignorance of the government administration for the food crisis in different parts of the CHT. "We had reminded the different concerned bodies of the government, but the authorities laughed at us and forgot what we said," says Dewan. He claimed that if the Regional Council (RC) had functioned properly according to the CHT Peace Accord, it could face the food crisis properly. He informs that the RC has asked for assistance from the donor community.

A source in the Ministry of the CHT Affairs informed that the ministry took this situation very seriously. Representatives of several donor agencies like UNDP and WFO had visited some of the affected areas in the CHT in the middle of April 2008. They chalked out a draft action plan.

The Problem behind the Problem: People suffering from the rat flood in the CHT were mostly *jumias*. Although *jum* (slash-and-burn cultivation) is a suitable agricultural method in hilly regions like CHT, it has never received any attention or support from the government. On the contrary, the government and some leading finance providers [ADB in particular] have always discouraged *jum* cultivation in the CHT. A successful *jum* depends on land-man ratio. The longer the fallow period, the better is the productivity of a *jum*.

The ratio of land-man that existed for an effective *jum* cultivation for centuries, has been seriously disturbed since the construction of the Kaptai Hydroelectricity project in the early 1960s. The influx of Bengali settlers sponsored by the state in the 1970s further disturbed land-man ratio. This drastically reduced the land available for *jum* cultivation. This led to drastic reduction in the fallow period from 15 to 20 years in the past to 2 to 3 years in the recent times. Consequently, the yield from *jum* has declined sharply. The bamboo flowering and rat flood makes the condition of *jum* and *jumia* worse.

Since the flowering of bamboo, at least four persons died reportedly of starvation. However, Jagot Jyoti Chakma, chairman of Rangamati District Council, said that he was unaware of those deaths.

Facing the Problem: As of May 2008 the government had sanctioned Tk.2.2 million (22 lacs) to combat the food shortages in different regions of the CHT. UNDP distributed a package of 20 kg rice, 1 kg salt, 1 kg dried shrimp powder, and two rat traps to 7,000 worst affected households. However, this was considered insufficient compared to the need, admitted Jagot Jyoti Chakma.

Kazal Talukdar informed, "We have trained the farmers to control the rats and instructed the dealers at the upazila level to stock a good amount of rodenticide. He said, "Our (government) attitude towards *jum* is quite unclear. We neither encourage *jum* nor discourage it."

Mizoram Situation: The government of the Indian state of Mizoram, had prepared a three-step action plan under a project, Bamboo Flowering and Famine Combat Scheme worth Rs.500 crore to combat the rat flood and its consequences. The actions were: (a) Early harvest of bamboos. The government has begun cutting down all bamboo plants to sell in different parts of India. (b) The state started a program to control rodents. Accordingly, Rs.1 was given for each rat killed. The tail of the rat had to be produced to collect the money. (c) Measures for regenerating bamboo plants that have been subjected to the harmful flowering.

In view of experience and the need for research and proper handling of the matter, universities in India have incorporated the bamboo and rat flood topic in related courses.

Conclusion: The flowering of bamboos in the CHT and the food shortage was not to end shortly. The local people knew, once bamboo started flowering, the rat flood would continue for at least three years. After the damages in 2007 due to rat flood, the *jumias* were afraid to engage in *jum* the following year.

A well-coordinated plan and concerted effort was necessary to tackle this situation. It is a fact that Bangladesh had no preparation to face the rat flood from bamboo flowering. Although the field level workers had cautioned the authorities in the center about the danger of the rat flood, they turned a deaf ear and remained silent. It is just an example of how the voices of the local people are ignored. However, *better late than never* was the only hope for the hill dwellers.

Source: Abridged. *Earth Touch* July 2008.

Traditional Use of Medicinal Plants in the Chittagong Hill Tracts

Sarder Nasir Uddin

Introduction

Plants used by the ethnic communities in curing different diseases are known as ethno-medicinal plants. Those plants have been used for curing different diseases from the beginning of human creation. Even in the present day, scientists have synthesized many modern medicines from medicinal plants (for example, quinine from sincona, morphine from aphim, atropine from balladona, etc.). During the last few decades scientists all over the world have been paying much attention to the studies in a newer branch of science called 'ethnobotany', especially to indigenous medicine or ethno-medicine. Since the 1980s this study has been intensified. Many countries of the world like India, Sri Lanka, Thailand, China, and Japan have developed a number of herbal medicines used in modern treatment by using medicinal plants. These countries not only use those medicines but also export these to other countries. In comparison to these countries, Bangladesh is far behind.

Bangladesh has a vast reservoir of plant resources. Medicinal plants are one of the important natural resources of Bangladesh. Being a tropical country and having fertile soil, Bangladesh possesses a good number of medicinal plants in every corner. The total number of medicinal plant of Bangladesh is still unknown. It is expected that out of 5,000 plant species found in the country, about 1,000 contain medicinal properties (Mia, 1990). Yusuf *et al.* (1994) and Ghani (1998) recorded 550 and 449 medicinal plant species from Bangladesh respectively. Bangladesh is an over-populated developing country. More than 60 percent of her population is out of reach of modern treatment facilities. The people living in the rural areas use medicinal plants for their primary health care and for the cure of different diseases for ages. The Chittagong Hill Tracts possesses the largest tropical rain forest of Bangladesh, which contains a vast amount of medicinal plant species. The majority of the ethnic communities of the country also inhabit this area. Living close to nature, the ethnic communities are custodian of a unique traditional knowledge system and wisdom about ambient flora and fauna. All these ethnic communities possess certain knowledge of useful drugs to be extracted from the jungle produce of the district.

In the past, the hill people were averse to coming in for a medical treatment and in most cases preferred their own methods of treatment. This was in no way due to dislike or fear of this treatment but to the great inconvenience of going in and being treated at the hospitals. To convey a serious case to the hospital meant a considerable amount of inconvenience and a derangement of the daily routine to a hill family. Herbal medicines prepared by the herbal practitioners (*boidayas*) were the only reliance of treatment of the ethnic people for their primary health care due to their independent culture, the poor transport system in the remote hilly areas, and the lack of modern treatment facilities. In the past, people of this region were completely dependent on natural resources not only for the treatment of different diseases but also for their livelihood. However, in order to adjust themselves to the passage of time, the ethnic people are getting educated and choosing various alternative means of livelihood instead of their traditional occupations like shifting (*jum*) cultivation, fishing, hunting etc. Many ethnic people are quitting their forefathers' occupation and are getting involved in different developmental works of the region. Nowadays many ethnic patients do not prefer to go to the herbal practitioners (*boidayas*) on account of the development of a good communication system and easily available modern treatment facilities in the districts.

The ethnic people have been widely and effectively using herbal medicines as remedy of various diseases in the region for many generations. Since most of these ethnic communities do not have their own scripts, the information about prescriptions, pharmacology, attitude towards diseases, diagnosis etc., of the age-old medicines are lying unclaimed. As the knowledge of those *boidayas* is mostly unrecorded, this is being lost when they quit their jobs. On the other hand, the destruction and alteration of natural habitats in the hill districts made many plant species threatened or endangered in the wilds. Hence, this valuable indigenous wealth of plant species of medicinal value including the knowledge of their uses in the Chittagong Hill Tracts (Rangamati, Khagrachari, and Bandarban) was felt to be thoroughly surveyed, inventoried, and identified. Bangladesh National Herbarium, with the sponsorship of the Ministry of Chittagong Hill Tracts Affairs, formulated a project titled, "Survey and identification of the medicinal plants of Chittagong Hill Tracts, exploration of their medicinal properties and uses, and publication of a monograph" in 2003. The main aims of the project were to (i) record the unrecorded traditional knowledge of medicinal plants and their uses, (ii) identify unknown medicinal plants and their status, (iii) collect and store those specimens in the Bangladesh National Herbarium as voucher specimens, and (iv) publish a pictorial monograph. The present study was conducted under that project.

Previous Works

A good number of researchers have contributed much towards the medicinal plants of the country viz. Hassan and Khan, 1986; Mia and Huq, 1988; Yusuf et al., 1994; Hassan and Khan, 1996; Chowdhury et al., 1996; Alam et al., 1996; Ghani, 1998 & 2002; Uddin et al., 2001; Khan et al., 2002. However, their contribution towards knowledge of the ethno-medicinal plants of the Chittagong Hill Tracts remains far behind. Natural scientists have given little attention so far to natural resources as well as to the biodiversity of the CHTs district due to the unstable political situation, this among other reasons. Heinig (1925) is the first person to prepare a list of plants of Chittagong Collectorate and Hill Tracts but he did not focus on uses of those plants. Though the works of Khisa (1996), Alam (1992) and Uddin et al., (2004) contributed much towards the ethno-medicinal plants of the CHTs region, a comprehensive work on the traditional knowledge and ethno-medicinal plant is still far to be completed.

Methodology

Extensive, careful, and planned field survey was carried out for two years among the 11 ethnic communities living in the Chittagong Hill Tracts with special concentration of the ethnic communities, which are believed to be isolated from the urban population to some extent. The knowledge about medicinal plants and their uses in the treatment of various diseases among the ethnic communities is often rather specialized and limited to a few members of a community who are recognized as *boidayas*. These persons are generally the most respectable and rather indispensable members in the ethnic societies. Those people were targeted for collection of information during the survey. In general, the *boidayas* treat all kinds of illness but some are specialized in specific illnesses. Moreover, some treats only children, or women or the aged and some practitioners have inherited knowledge of certain special remedies.

Information was gathered by interviewing a total of 228 informants (in most cases *boidayas* were the informants) and voucher specimens of each medicinal plant were collected for preserving at the Bangladesh National Herbarium. A standard format questionnaire was prepared for recording data. During the course of study the author along with project personnel visited *boidayas* repeatedly to collect data from them. They visited those informants' houses to record the diseases treated by them, medicine making procedure and their doses. Then the informants were requested to identify the plant. Alternately, a particular plant was picked up and queries were made as to how it was useful for them. When discussion on one plant was over, a second plant was taken up, and so on.

There is often a feeling among the ethnic communities that outsiders consider their customs to be funny or even absurd and there is a concomitant reluctance to expose themselves to casual visitors. Though they are usually not very unwilling to disclose their knowledge about the use of plant wealth except for the medicinal plants yet they maintain secrecy about the use of certain medicines, such as medicine of refractive diseases, diseases of women, contraceptive, and herbs for causing abortion. There are two main reasons behind that secrecy: (i) they believe that the medicines will lose their healing power if too many people know about them, and (ii) if their neighbors learn the formula then they will compete in their business. To overcome those situations, the author established a friendly relationship with them before asking about any medicinal plants and their uses. Secondly, emphasis was given to make them understand that the information would be preserved for the benefit of their future generations and their neighbors would not be told about the information given in confidence. In many instances, it was found that after understanding the genuine purpose of the study, the informants brought plants or plant parts to the field camp and narrated the use of their own initiative.

Once the information on a particular plant was taken as reliable after repeated verification, its local name and use were recorded. Details about the part utilized in preparation of the medicine, the ailments, the preparation, doses, and prescription were also recorded.

Major Findings

The study has found that the ethnic *boidayas* do the treatment of 301 human ailments with 2,295 different prescriptions. It has also found that they treat common diseases like cold and fever as well as some incurable diseases like cancer. Sometimes they use a single plant species for the treatment of a single disease such as anal fistula, beriberi, and blood cancer. At other times, they use a number of plant species for the treatment of other disease such as tuberculosis (50 species) and lipoma (45 species). In some instances, there is only a single prescription for the treatment of a single disease such as duodenal ulcer, haematurrhoea, and mastitis. In other cases, a number of prescriptions are available for the treatment of a single disease such as rheumatism (89 prescriptions), jaundice (80 prescriptions), and stomachache (67 prescriptions). However, in a majority of cases, they use more than one formula for the treatment of a single disease. It has also found that different *boidayas* use different formulas to treat the same disease. A total of 700 plants species are known to be used as medicines by the ethnic communities living in the CHT. Among these plant species some plants grow throughout the country (e.g. *Arundo donax, Blumea lacera, Calotropis gigantea, Clerodendrum viscosum, Commelina benghalensis, Cuscuta reflexa, Eupatorium odoratum, Heliotropium indicum, Leucas aspera, Ludwigia perennis, Microcos paniculata, Mikania cordata, Mimosa pudica, Scoparia dulcis, Spilanthes calva, Trewia nudiflora, Vernonia patula* and others grow only in the CHT (e.g. *Phobe lanceolata, Wallichia densiflora, Baccaurea lanceolata, Blumea clarkei, Cadaba indica, Cheilanthes belangeri, Cheilanthes farinosa, Cynoglossum hellwigii, Doryopteris ludens, Mantisia spathulata, Stauranthera grandiflora, Vitis pallida*). Some plant species have also become threatened in nature (i.e. *Asparagus racemosus, Brownlowia elata, Curcuma amada, Dehaasia kurzii, Holigarna longifolia, Hydnocarpus kurzii, Licuala peltata, Mantisia spathulata, Ophiorrhiza villosa, Plumbago indica, Rubus moluccanus, Rubus hexagynus, Stemona tuberosa*). Many of these plant species (e.g. *Asparagus racemosus, Emblica officinalis, Rauvolfia serpentina* etc.) have been used in medicines of modern allopathic and homeopathic treatment. It is found that many plants used by the ethnic communities have important compounds, which have been used in modern medicines and cosmetics. Some of these plants could be cultivated for commercial purposes.

Commons Diseases among the Ethnic People

The common diseases among the ethnic people are biliousness, fever, colic, diarrhea, dysentery, dyspepsia, gingivitis (inflammation of gum), hepatic diseases, indigestion, influenza, malarial fever, pneumonia, skin diseases (e.g. ringworm and scabies), ulcer, venereal diseases and others. Diseases like paralysis, urticaria, jaundice, and conjunctivitis are not very uncommon. Some diseases are reported as specific to men or women or children in ethnic communities. The common female diseases are amenorrhea, anemia, menorrhagia, dysmenorrhea (painful menstruation), hysteria, mastitis, leucorrhoea, etc. The practice of abortion is not uncommon. On the other hand, the common diseases among the children are anemia, bronchitis, caries of teeth, cough and cold, dysentery, diarrhea, measles, harpies, itch, intestinal parasites, mumps, night blindness, otitis, scurvy, strangury, whooping cough, etc. Some of these diseases become lethal among the ethnic children with a combination of malnutrition, which is very common. Most of the ethnic people believe that the accidental inhalation of the cold air can make a person very sick. They also believe that complicated diseases and ailments such as typhoid, pneumonia, piles, goiter, and rheumatism originate from a spell or curse, to the evil spirits or exorcism, the violation of some Gods, or to the work of sorcerers, who are inimical.

Local Names of Plants: Indicator of Medicinal Properties

In general, the ethnic people give names to plants with known good or bad properties. The absence of any name to a plant is not the evidence of its uselessness. It is common that most of the *boidayas* have their own secret names for many of the medicinal plants to safeguard their specialized knowledge. Thus more than one local name has been attributed to a plant, which generally creates confusion to ascertain the actual plant species. They have some general views regarding medicinal plants. Those are: (a) plants with latex and bitter taste have medicinal value, (b) green plants before flowering are more effective as curative agents than the dried plants, (c) roots are considered more effective medicine than the aerial parts of the plants, (d) plants used for medicinal purpose in various diseases should be collected at different times as in the morning, midday or in the afternoon, (e) combination with the fruits of black pepper (*Piper nigrum*) the medicine shows more efficacies, (f) The barks in medicinal use should be taken from that side of the plant on which the rising sun shines and (g) the ethnic people have a belief that the violation of any taboo makes the medicine ineffective.

Ethnic Pharmacology

The preparation of drugs and medicines among the ethnic communities is really an art. Most of the ethnic medicines are prepared either from a single drug obtained from a single plant or plant part. The combination with other plants, animal organs, rocks, minerals, salt etc., in ethnic medicine is not uncommon. Practically there is a lack of standardization both in the case of weighing in measure and the preparation of drugs. These vary from one ethnic community to the other, even from one herbal practitioner to the other of the same community. The authors have tried to bring some uniformity among them. Different forms of drugs used by the *boidayas* for curing various diseases are given below:

Fresh juice: The fresh plants or plant parts are cut into pieces and are crushed either by mortar like stone or in between the two palms. Then the extract is squeezed out. It is believed that fresh juices has the highest potency and is administered orally or applied externally. Fresh juice should be prepared every time before use.

Paste: Both the fresh and dried materials, after cleaning, are made into paste on a stone and are used both orally and applied externally. The paste is usually used fresh or can be preserved for 24 hours by adding honey or common salt.

Pill: The powder of the dried plants materials or paste of the fresh plant materials mixed with some edible gums, honey, maize or rice powder, which act as binding materials, are made into small round pills and then dried in the sun. The pills should be taken orally and can be preserved for a long time.

Extract: Both the fresh and/or dried materials are boiled in a wide mouth earthen pot in water or alcohol (1:20 ratio) and are reduced to about one third of the original volume. The extract is used internally and may be preserved for 2-3 weeks.

Infusion: The dry materials, after cleaning, are cut into pieces and soaked in drinking water about 1:6 ratio overnight. The decanted water is used

(top row from left) *Abroma augusta* L. & *Phyllanthus emblica* L. (2nd row) *Cath ranthus roseus (L.)* G. Don & *Centella asiatica* (L.)Urban (3rd row) *Gloriosa supe ba* L. & *Mesua ferrea* L. (4th row) *Premna esculenta* Roxb. & *Rauvolfia serpenti* (L.) Benth. *ex* Kurz.

op row from left) *Bombax ceiba* Burm f. & *Ipomoea mauritiana* Jacq. (2nd row) *ubus moluccanus* L. & *Saraca asoca* (Roxb.) W.J.de Wilde (3rd row) *Stemona tube-sa* Lour. & *Terminalia arjuna* (Roxb. *ex* DC.) Wt. & Arn. (4th) *Terminalia citrina* Gaertn.) Roxb. *ex* Fleming & *Thunbergia grandiflora* Roxb. © Sarder Nasir Uddin

as medicine and taken orally. Sometimes a pinch of salt or a tea spoon of honey is added to it. The infusion is used internally and cannot be preserved more than one day.

Decoction: Both the fresh and/or dried materials are boiled in wide mouth earthen pot in water (1:10 ratio) and are reduced to about one fourth of the original volume. Sometimes a pinch of salt or tea spoon of honey is added in it. The decoction is used internally and may be preserved for 2-3 weeks.

Mixture: This is either the combination of the powders of dried materials or that of the decoctions and extracts of plant materials in a certain ratio. The mixture is used internally and may be preserved for 2-3 weeks in a tight mouth jar.

Powder: The dried materials are pounded and made into fine powders. The powders are used internally and preserved for a long time. It is believed that the medicine gets more potency by storage for a long time.

Syrup: The powdered dried raw materials are soaked in drinking water and put in an earthen pot covered with lids for 3-4 days. The contained materials are shaken for about 30 minutes every day. The decanted liquid is then mixed with honey (4:1 ratio) is considered as syrup. The syrup should be taken orally and can be preserved for several months.

Fomentation: The fresh leaves, mainly of latex bearing plants, are put on fire for about few minutes after adding a little oil. Then the leaves are allowed for cooling for about 1-2 minutes and applied externally on the affected parts of the body.

Massage balm: The dried plant and animal produces are mixed with oil or butter or ghee by stirring for hours. Then it is boiled for about 30 minutes to one hour. The massage balm should be used externally and can be preserved for a long time in normal temperature.

Fume: The fume is created by burning the dried material or boiling dried or fresh material in water. The fume is inhaled by covering the part of the body with cloth.

Ash: The dried plant materials are put into fire and allowed for full combustion. After cooling, the ash is taken out and applied with oil or water.

Plaster: The pastes of fresh plant materials in a specific ratio; sometimes lime is added to it.

Prescriptions and Doses

In general, the dose prescribed for each medication is for adults. It is either reduced or increased proportionately with the age of patients. Different forms of ethnic medicines are prescribed for taking orally and applying externally depending on the disease type. In general, the ethnic people take the medicine either with fresh cold drinking water or with country liquor, rice beer, or with honey as advised by the *boidaya*. Most of the ethnic medicines are prepared in combination with the fruits of black pepper. Sometimes they also add ginger. They believe that addition of pepper seeds in the drug make the medicine more therapeutically effective. Probably pepper and ginger are used as bio-assimilating agents. Duration of application and doses of a medicine depend on the type of the disease. Some medicines are advised for taking for one to two times (i.e. diarrhea, stomachache, headache). In case of common diseases like fever and dysentery medicines are advised for taking three to seven days. In case of acute and refract diseases like rheumatism, paralysis, cancer, and impotence medicines are advised for taking from one month to six months or for longer period.

Problems with Ethno-medicinal Treatment

The increasing demand for herbal treatment throughout the world has also increased the trade in medicinal plants and herbal medicines. Though it is a prospective field, the practice of herbal treatment in the CHT has been facing serious problems. Some of them are:

(i) The knowledge about herbal medicines is apparently non-transferable. Many *boidayas* are unwilling to disseminate their knowledge to others. On the other hand, there is no institutional education system or even printed books on traditional herbal medicine for teaching interested people. Besides these, the linguistic problem and the complex manufacturing and the difficult application procedure of herbal medicine in different diseases has made the subject least popular.

(ii) The unavailability of important medicinal plants in nature due to over exploitation and destruction of forests. The *boidayas* find it difficult to prepare medicines.

(iii) The patients' lack of confidence in *boidayas*, because many do not have proper knowledge about diseases and herbal medicines. The symptom-based herbal treatment might cause death or great sufferings of the patients, if diagnosis of a disease goes wrong.

(iv) Nowadays modern allopathic treatment has become available to the people even in the remote areas thanks to the development of a good communication system. Hence, people prefer to call in an allopathic doctor instead of local herbal practitioners. In some cases, herbal treatment is found more costly than allopathic treatment. So, people do not want to take those expensive herbal medicines.

(v) The number of herbal practitioners (*boidayas*) has been declining because of their poor financial condition. The survey has found that only a few *boidayas* are doing well in their profession in the hill districts. In many cases, patients want the cure first before paying any money; in some cases they don't pay any money at all. Hence, most of the *boidayas* cannot afford to live with their own business.

(vi) Sometimes people are afraid of *boidayas* because a few of them are vindictive and do harm to the people by using poisonous plants. Besides herbal treatments, some *boidayas* may cast a spell on the people to harm them.

Conservation of Medicinal Plants

Only few selected medicinal plants are cultivated in Bangladesh. A large portion of medicinal plants is being collected from the wild or imported. The demand for medicinal plant as well as herbal medicine has been increasing sharply and that has lured the local medicinal plant collectors. They usually collect medicinal plants indiscriminately and in an unsustainable way from the wild that has led to the rapid depletion of a number of medicinal plant species. Some have become threatened and others have become vulnerable. For example, in the past *Rauvolfia serpentina* and *Asparagus racemosus*, two very important medicinal plants, were collected from the wild and exported so widely that both plants are threatened to extinction. Those two plants have been listed in the Red Data Book of Vascular Plants of Bangladesh. Hence it is urgently needed to conserve the medicinal plant species. Otherwise many species having potential medicinal value will be extinct in the nature in the near future.

Recommendations

The importance of medicinal plants is universal. More than 70 percent of the rural people of Bangladesh depend on medicinal plants directly or indirectly for their primary health care and treatment for different diseases but a large number of medicinal plant species and their uses are yet to be known. A large portion of medicinal plants is generally collected from the wild and thus has reduced many plant populations in nature. A large number of people are engaged in collecting medicinal plants, the practice of herbal medicines and pharmaceutical industries. A large number of important and commercially valuable medicinal plant species can be grown in Bangladesh for producing herbal medicines. Therefore medicinal plants and herbal medicines could be a very important sector. The following are some recommendations to make the sector popular and profitable.

Both *in-situ* and *ex-situ* conservation measures should be taken for the conservation of the threatened plant species. The local community should be involved in this activity. It is well known that most of the herbal practitioners may collect plants from the wild without leaving individuals. Hence, free access into the forests to exploit naturally growing medicinal plants should be restricted. The policies pertaining to the collection of medicinal plants and other non-timber forest products should be revised.

The cultivation of medicinal plants should be done in a systematic way and according to the demand of the market. Training should be given to the farmers about the cultivation, processing and harvesting of medicinal plants.

The quality of herbal medicine must be ensured by standardization in post harvest processing to gain the confidence of the customers. For this purpose, research should be carried out on herbal medicines and on the scientific standardization of medicines.

The safety and efficacy of the herbal medicines are to be scientifically evaluated. Although it is very often said that natural medicine is safe as it is being used for ages, the incidence of adulteration and side effects are numerous especially on long-term uses.

The standardization methodology of the herbal medicine preparations by using chemical and spectroscopic fingerprinting is to be developed. For quality assurance of the raw materials and final products is simple but latest scientific procedures can be applied.

More research should conducted to develop new medicines, method of cultivation, harvesting techniques, processing techniques for packaging and transportation, etc. For these purposes inter-institutional co-ordination, co-operation and linkages should be developed.

The government should adopt policies to promote the sector by giving training to the cultivators, creating markets, storing and processing facilities, and developing indigenous knowledge based herbal medicine industries.

References

Alam M.K. 1992. Medical ethnobotany of Marma tribe of Bangladesh. Economic Botany 46 (3): 330-335.

Alam, M.K., Chowdhury, J. and Hassan M.A. 1996. Some folk formularies from Bangladesh. J. Life Sci. 8(1): 49-63.

Chowdhury, J., Alam, M.K. and Hassan M.A. 1996. Some folk formularies against dysentery and diarrhoeain Bangladesh. J. Econ. Taxn. Bot. Additional Series 12: Scientific Publisher, Jodhpur India. Pp. 20-23.

Ghani, A. 2002. Veshaj Ousad. Bangla Academy, Dhaka.

Ghani, A. 1998. Medicinal Plants of Bangladesh: Chemical Constituents and Uses. Asiatic Societies of Bangladesh. Dhaka.

Hassan, M.A. and Khan, M. S. 1986. Ethnobotanical record in Bangladesh-1, Plant used for healing fracture bones, J. Asiatic Soc. Bangladesh (Sci.) 12: (1 & 2): 33-39.

Hassan, M.A. and Khan, M. S. 1996. Ethnobotanical record in Bangladesh-2, Plant used for healing cuts and wounds. Bangladesh J. Plant Taxon. 3(2): 49-52.

Heinig, R.L. 1925. List of plants of Chittagong Collectorate and Hill Tracts. The Bengal Government Branch Press, Darjeeling. pp. 1-78.

Khan, M.S., Hassan, M.A. & Uddin, M.Z. 2002. Ethnobotanical survey in Rema-Kalenga wild life sanctuary (Habigonj) in Bangladesh. Bangladesh J. Plant Taxon. 9(1): 51-60.

Khisa, B. 1996. Chakma Talik Chikitsha. Rajbon Bihar, Rajpari, Rangamati.

Mia, M.M.K. 1990. Bangladesh flora as a potential source of medicinal plant and its conservation strategies. In: Ghani, A. (ed), Traditional Medicine, Jahangirnagar University, Savar, Dhaka, pp. 73-79.

Mia, M.M.K. & Huq, A.M. 1988. A preliminary ethnobotanical survey in the Jointiapur, Tamabil and Jaflong area, Sylhet. Bull. 3, pp. 1-10. Bangladesh

National Herbarium, Dhaka.

Uddin, M.Z., Khan, M.S. & Hassan, M.A. 2001. Ethnobotanical plant records of Kalenga forest range (Habigonj), Bangladesh for malaria, jaundice, diarrhea and dysentery. Bangladesh J. Plant Taxon. 8(1): 101-104.

Uddin, S. N., Uddin, M. Z., Hassan, M. A. & Rahman, M. M. 2004. Preliminary ethnomedicinal plant survey in Khagrachari district, Bangladesh. Bangladesh J. Plant Taxon. 11(2): 39-48.

Yusuf, M., Choudhury, J.U., Wahab, M.A., Begum, J. 1994. Medicinal Plants of Bangladesh, BCSIR, Dhaka.

Adhatoda vasica L.   © Sarder Nasiruddin

The Chittagong Hill Tracts: The Haven of Wildlife Under Severe Threat

M. Monirul H. Khan

Introduction

The diversity of life on earth that we see today did not originate suddenly. This is an outcome of a very slow evolutionary process that took 3.5 million years from the origin of life on earth. Very few areas on earth are blessed with extremely high diversity of life forms–from charismatic mega-fauna to microscopic organisms–together with high endemism, i.e. the species that are not found elsewhere. Scientists have designated these areas as 'biodiversity hotspots', which demand the highest priority for conservation. Conservation International (an international organization engaged in biodiversity conservation) has identified a total of 34 such hotspots on earth. One of them is Indo-Burma Biodiversity Hotspot (Conservation International 2012), which has its western end in the Chittagong Hill Tracts (CHT) of Bangladesh. This is the only hotspot that is shared by Bangladesh. The Indo-Burma Biodiversity Hotspot encompasses more than two million square kilometers of tropical Asia east of the Ganges-Brahmaputra lowlands. This hotspot harbors a total of 13,500 species of plants, 433 mammals, 1,266 birds, 522 reptiles, 286 amphibians, and 1,262 freshwater fish, with a lot of species endemic to this hotspot, and many species are yet to be discovered. Notably, six large mammal species have been discovered from the Indo-Burma biodiversity hotspot in the last 12 years. The area supports probably the highest diversity (53 species) of freshwater turtles in the world.

The inclusion of the CHT in one of the 34 global biodiversity hotspots has made the area globally and nationally important for biodiversity conservation. This is the most biodiverse part of Bangladesh where we still have a number of charismatic megafauna in the wild. This is the area where scientists expect to discover species new to Bangladesh, or even new to science. Sadly, however, the wildlife and their unique habitats in the CHT are vanishing rapidly.

With an area of about 13,184 km², the CHT is situated in the southeast of Bangladesh along the borders of Mizoram and Tripura states of India, and Myanmar. The area is divided into three administrative districts, viz., Khagrachhari, Rangamati and Bandarban. It is basically a hilly landscape with evergreen forests, bamboos and grasslands, but between the hills it embraces fertile valleys and riparian forests along Kassalong, Myani, Karnaphuli, Sangu and Matamuhuri river valleys, making the area a diverse habitat for various forms of wildlife. The temperature normally varies between 20 and 35°C during winter and summer; humidity is very high during the rainy season and the annual rainfall varies between 2,540 and 3,810 mm. The dry and cool season is November to March followed by the pre-monsoon of April and May, and a very hot and humid monsoon from June to October. The humid environment is good for the growth of plants and insects.

Wildlife Habitats

The reason why the CHT is exceptionally rich in wildlife is the existence of diverse habitats. The habitat diversity is the combined result of altitudinal variation and diverse vegetation. Originally, luxuriant evergreen and mixed evergreen forests characterized by a continuous high canopy covered most of the CHT. Bamboo groves, bushy plants, reeds and grasses covered rest of the CHT. The area also has a number of rivers and streams as well as wetlands and lakes, making the area suitable for water-loving species. Once only a small proportion of the land was occupied by the villages and by *jum* (shifting) cultivation by the people of the ethnic communities. Today, the land is not only taking the load of rapidly growing ethnic people, but also the settlers from the plains land, claiming most of the forests and other wilderness areas for habitation and agriculture, together with monoculture plantations for timber and rubber.

The CHT, however, still has relatively large tracts of natural forests in Kassalong Reserved Forest and Sangu Wildlife Sanctuary, and some smaller patches in Pablakhali Wildlife Sanctuary, Kaptai National Park, Reinkhyong Reserved Forest, etc. The giant Kaptai Lake, created due to the construction of a dam on the Karnaphuli river that was completed in 1963 for the production of hydro-electricity, caused the destruction and inundation of riparian forests, but now serves as a good habitat for waterbirds, reptiles and amphibians. Moreover, there are two huge natural lakes, viz. Boga lake and Reinkhyong lake. Today, most of the natural vegetations in the CHT are bamboo, bush, and reeds, offering habitats for many species unique to these vegetation types. The hills that remain abandoned for a few years after *jum* cultivation (to naturally refresh the soil fertility before the next round of cultivation) are also used by some forms of wildlife like rodents, warblers, babblers and lizards.

Other than the macro-habitats, there are some unique micro-habitats offered by caves, gorges, rocky areas, waterfalls, and narrow streams that are essential for some species. The caves in Boga, Rumna and Alikadam in Bandarban, and the gorge near Remakri, Bandarban, are ideal roosting places for many small forest-bats during the day. Similarly, the waterfalls and adjacent streams at Baklai, Jadipai, Rumna and Poamuhuri in Bandarban, and Reinkhyong *and* Sijok *in* Rangamati, offer unique habitats for some frogs, lizards and turtles.

Status of Wildlife

The most prominent wildlife species in the CHT is the Asian Elephant (*Elephas maximus*), which is one of the largest living animals on earth and a globally Endangered species (IUCN 2011). It is estimated that about 200 resident elephants occur in Bangladesh, of which about 130 occur in the CHT. The main strongholds of elephants are Eastern Rangamati, Southern Bandarban and Northeastern Khagrachhari. Due to their giant size elephants require a huge quantity of plant food every day. Since there are crop fields in and around the elephants' ranges, they are often attracted to and raid the crops. Since the crop raiding is more gainful than foraging for scattered and poorly nutritious food in the wild, elephants will raid crops whether or not there is food in the wild. Elephants also destroy human houses in search of stored crops that they can smell from far away. Moreover, the elephant corridors that they use for many years for their long movements are often blocked by human habitation. Therefore, a severe conflict between elephants and people exist in the CHT, causing death and injury to people and elephants. Interestingly, relatively less ethnic people are killed by elephants than the settlers, because the ethnic people have better knowledge of elephant behavior and movement pattern, and they

(left) Hoolock Gibbon pair foraging in the upper canopy of Kaptai National Park. (right) Clouded Leopard is a secretive carnivore that often climbs the trees in search of prey animals.   © Monirul Khan

try not to block the corridors.

The only wild bovid that still exists in the wild in Bangladesh is Gaur (*Bos gaurus*), although until recently it was thought that Gaur has gone extinct in Bangladesh and the occasional reports of Gaurs are vagrants coming from the neighboring countries. Recently local people (2011) have confirmed the occurrence of at least two small resident populations in Kassalong Reserved Forest, Rangamati (in Vulongtoli Mon and in Betling). Moreover, the local people occasionally see Gaur in Sangu Wildlife Sanctuary, Bandarban, but I am not sure whether they were residents or vagrants. In the CHT people have domesticated Gaur, which is locally called Gayal. People have also successfully produced a hybrid between Gayal and the ordinary cow, which is called Tongoru.

All the eight species of greater and lesser cats that occur in Bangladesh are known to occur in the CHT. Many people don't know that the Tiger (*Panthera tigris*) occurs not only in the Sundarbans, but also in the CHT. It is rarely sighted in Kassalong Reserved Forest and Sangu Wildlife Sanctuary,

(left) Red-headed Trogon, one of the most beautiful birds in the forests of the CHT. (top right) White-lipped Pit Viper, a common poisonous snake of the CHT. (bottom right) Asiatic Black Bear, rarely sighted in the CHT   © Monirul Khan

but it is not known whether they are residents or vagrants. I have heard the stories of recent tiger sightings (at least two sightings in 2011) from the local ethnic people. Two other greater cats, i.e. Leopard (*Panthera pardus*) and Clouded Leopard (*Neofelis nebulosa*), are also sighted rarely. Since the Clouded Leopard often roosts on trees and forages at night it is difficult to see, but in 2010 I saw their fresh pugmarks in Kaptai National Park, Rangamati. Lesser cats are more abundant than their greater cousins, but in Bangladesh the Asiatic Golden Cat (*Felis temminckii*) is rare and it is restricted only to the CHT, and there is no recent sighting of the elusive Marbled Cat (*Felis marmorata*).

At least two species of bears occur in the CHT, i.e. Asiatic Black Bear (*Ursus thibetanus*) and Sun Bear (*Helarctos malayanus*), but the latter species is extremely rare. The habitat of the CHT is ideal for bears and there is plenty of food for them, yet they are attracted to the *jum* for maize and other crops, and are killed by locally made guns or trapped by leg-traps and beaten to death. I saw the paws and other body parts of bears in the CHT. Bears are also killed, because some people believe that its gall bladder has medicinal value.

Eight out of ten species of primates of Bangladesh are known to occur in the CHT. Hoolock Gibbon (*Hylobates hoolock*) is the only ape in South Asia. Phylogenetically, this species has close relationship with humans, which are supported not only by its lacking a tail, but also by its complex behavior and made of parental care. Male and female gibbons make partnership for lifetime, unless one partner dies. Rhesus Macaque (*Macaca mulatta*) is the commonest of primates. It is well adapted to diverse habitat conditions and food, and is known to be very intelligent. As a consequence, Rhesus Macaque has a healthy population in the CHT despite the fact that the species is frequently hunted by the local hunters.

Three species of deer occur in the CHT of which Hog Deer (*Axis porcinus*) has long been treated as 'extinct' in Bangladesh (Khan 1982, IUCN-Bangladesh 2000) until I reported a juvenile male that was captured in 2002 by the local trappers in Guimara, Khagrachhari (Khan 2004). Later on, an adult male was trapped in Tabalchhari, Khagrachhari. The largest deer of South Asia is Sambar (*Cervus unicolor*), which weighs 225-320 kg. It occurs in the CHT, preferably in and around the areas where some forest patches still exist.

Among the smaller mammals there are many notable species like Hodgson's Giant Flying Squirrel (*Petaurista magnificus*), Chinese Pangolin (*Manis pentadactyla*), Lesser Bamboo Rat (*Cannomys badius*), etc. Hodgson's Giant Flying Squirrel cannot really fly, but glides from tree to tree. It is a nocturnal species. I was lucky to see one gliding from tree to tree in Kaptai National Park, Rangamati, one evening. It was really amazing how it manages to glide over the long gaps between the trees. Chinese Pangolin is interesting to many people, because it has overlapping scales that resembles that of fish. However, these scales are actually modified hairs. The Bamboo Rat is a cute rodent that spends most of its life underground. It feeds on roots and tubers while digging underground tunnels. Moreover, a wide variety of rodents, bats, and other small mammals exist in the CHT.

The evergreen forests of the CHT are characterized by having high canopies, which are ideal habitats for hornbills. Two large hornbills, i.e. Great Hornbill (*Buceros bicornis*) and Wreathed Hornbill (*Aceros undulatus*), are extremely rare and occur in the eastern CHT, which do not occur elsewhere in the country. I saw Great Hornbill in Sangu Wildlife Sanctuary and in Kassalong Reserved Forest. Oriental Pied Hornbill (*Anthracoceros albirostris*), however, is common and also occurs in the forests of Greater Sylhet.

A number of large ground-dwelling birds occur in the CHT, of which there is no recent report of Green Peafowl (*Pavo muticus*), but it might still occur in remote forests of the CHT. The beautiful Grey Peacock Pheasant (*Polyplectron bicalcaratum*) is very rare and secretive, but I was lucky to see it in Kaptai National Park. Other ground-dwelling birds like Red Junglefowl (*Gallus gallus*; ancestor of all varieties of domestic chickens), Kalij Pheasant (*Lophura leucomelanos*), White-cheeked Partridge (*Arborophila atrogularis*) and Barred Buttonquail (*Turnix suscitator*) are rather common.

The only sighting of Indian Courser (*Cursorius coromandelicus*) in Bangladesh occurred on the shore of the Kaptai lake in 1998. Most of the birds of evergreen forests of the CHT are vividly colored and would easily attract any birdwatcher. Some of the brightly colored birds are Red-headed Trogon (*Harpactes erythrocephalus*), Blue-bearded Bee-eater (*Nyctyornis athertoni*), pittas and minivets.

Diverse species of reptiles and amphibians occur in diverse habitats of the CHT. Two species of pythons, viz. Burmese Python (*Python molurus*) and Reticulated Python (*Python reticulatus*), occur in the CHT, of which the latter is rare. On three occasions I saw them in Sangu Wildlife Sanctuary, Bandarban. Both species of python grow very large, but the Reticulated Python holds the largest snake's record on earth (7.3 m long). Both species are hunted for meat by the local ethnic communities of the CHT. Among the poisonous snakes, White-lipped Pit Viper (*Trimeresurus albolabris*) is the commonest and King Cobra (*Ophiophagus hannah*) is the largest, which is also the largest poisonous snake on earth (3.8 m long). One interesting lizard is Spotted Flying Lizard (*Draco maculatus*), which has adapted to the arboreal habitats of evergreen forests, gliding from tree to tree in search of ants and other insects. Its gliding membrane is very colorful, but the color is visible only when the membrane is expanded during the glide. Among the frogs there are tree frogs and land-dwelling frogs. The large Tree Frog (*Rhacophorus maximus*) is a giant tree-dweller that is extremely rare. There are only two sightings of it in Bangladesh, of which I saw one in Ruma, Bandarban.

Wildlife Vanished

We have already lost at least 13 species of large and charismatic wildlife species from Bangladesh of which at least five species used to occur in the

CHT (Khan 2008). These are Banteng (*Bos javanicus*), Wild Water Buffalo (*Bubalus arnee*), Sumatran Rhinoceros (*Dicerorhinus sumatrensis*), Indian Rhinoceros (*Rhinoceros unicornis*) and Swamp Deer (*Cervus duvaucelii*). It is feared that the total number of wildlife species that went extinct from the CHT and from Bangladesh is higher, but we need more time and surveys to confirm their absence. Banteng used to occur in the CHT until the 1930s and the Wild Water Buffalo used to occur until at least 1940s. Two species of rhinoceros used to occur in the CHT until the end of the 19th century. The Swamp Deer existed in the CHT until the 1950s.

The wildlife species that are feared exterminated from the CHT, but the fact awaiting confirmation are Sloth Bear (*Melursus ursinus*), Green Peafowl (*Pavo muticus*), Rufous-necked Hornbill (*Aceros nipalensis*) and White-winged Duck (*Cairina scutulata*).

Wildlife New for Bangladesh

The CHT is the least explored area of the country where very few surveys on wildlife have been conducted. In recent years a number of wildlife species have been reported from the area not listed in the relevant checklists of Bangladesh (Khan 2008, Siddiqui *et al.* 2008, Ahmed *et al.* 2009, Kabir *et al.* 2009, Khan 2010). From November 2010 to November 2011, I have recorded two species of small mammals, one species of turtle and two species of frogs in the CHT, which are new for Bangladesh. In recent years, others have reported one bird and one turtle species from the CHT, which are new for Bangladesh.

The two small mammal species that I have recorded are Himalayan Striped Squirrel (*Tamiops maclellandi*) and Least Leaf-nosed Bat (*Hipposideros cineraceus*). The former is a tiny arboreal squirrel that was found in November 2010 in a healthy evergreen forest patch in Theikkang, Ruma, Bandarban, and the latter is a tiny bat that was found in November 2010 in a natural cave in Boga, Ruma, Bandarban. I found Asian Softshell Turtle (*Amyda cartilaginea*), in November 2011 in Remakri Khal, Thanci, Bandarban. Among two species of new frogs that I found, Anderson's Bush Frog (*Philautus andersoni*) was found in June and July 2010 in the evergreen forest's undergrowth of the Kaptai National Park, Rangamati, and the Nicobarese Frog (*Hylarana nicobariensis*) was found in November 2010 in a small ditch at the top of a hill in Moyu, Roangchhari, Bandarban. Moreover, the Mountain Hawk Eagle (*Spizaetus nipalensis*) was sighted in November 2009 in Belaichhari, Rangamati (R. Halder, personal communication), which was the first sighting of the species in Bangladesh. In November 2010, I saw it in Keokradong Range, Ruma, Bandarban, for the second time. In December 2011, a shell of recently hunted Keeled Box Turtle (*Cuora mouhotti*) was photographed from the ethnic Mro hunters in Sangu Wildlife Sanctuary, Bandarban (S. C. Rahman, personal communication), which was the first record of the species in Bangladesh.

Threats to Wildlife

The biggest threat to wildlife of the CHT is habitat loss due to rapid deforestation and overexploitation of natural resources, mass settlement of plains land people, expansion of monoculture plantations, and unplanned expansion of *jum* cultivation. Other threats include lack of strategically chosen protected areas, lack of law enforcement, overhunting of wildlife, insufficient animal protein for people from cultivated fish or domestic chicken and livestock, and lack of awareness.

To meet the demand of timber and firewood of the booming human population, the hardwood trees and other types of non-timber forest products are legally and illegally overexploited in the CHT. Poor law and order enforcement and corruption among the managers and law enforcing agencies have made it easy to clear the natural timber trees as well as planted teak and other trees, together with bamboo, cane and other forest products. As a consequence, the wildlife species lose the habitat and either die or disperse elsewhere. Although several species of naturally occurring bamboo are harvested in large quantities, the bamboo groves are less severely affected, because it can regenerate rapidly. The mass death of bamboo due to flowering and fruiting, however, temporarily destroys the habitats of many wildlife species.

The traditional agricultural practice in the CHT is *jum* or shifting cultivation. A few decades ago the *jum* cultivation was not a threat to wildlife habitats, because there were very few people in vast tracts of hills, but today the hill population has increased and, to some extent, commercialization has influenced the hill people to produce crops not only for their own consumption, but for sale.

In 1980s the Government of Bangladesh took a political decision to shift and settle the poor landless people from plains districts to the CHT. In most of the cases these people were settled in the Government-owned reserved forests by clearing the areas and changing the status of the lands as de-reserved areas. The coming of plains land people who had no traditional knowledge of leading a life in hills became disastrous for the forests and wildlife of their surroundings. Today, no good natural forest exists in and around the settlement areas.

In the early 20th century a British forester brought seeds/seedlings of teak from Myanmar (Burma) and planted them in Kaptai, Rangamati, as an experiment. Following the success of this experiment, huge areas of hills had been cleared for teak plantation. The mature teak trees of the earliest plantation still exist in some areas of the CHT, which are a major attraction to the illegal loggers. In the last few decades not only teak, but also rapid-growing exotic species of timber trees (acacia, eucalyptus, etc.) and rubber trees were planted, further shrinking the forests and other natural wilderness areas that served as habitats for wildlife.

Other than habitat loss, there are many other direct and indirect threats to wildlife. There are only three formal protected areas in the CHT (Pablakhali Wildlife Sanctuary, Kaptai National Park and Sangu Wildlife Sanctuary) covering only 3.8 per cent of the CHT. Many rich forests, bamboo groves, grasslands and other vegetation types as well as freshwater wetlands of the CHT are not included in the protected area network. Even the existing protected areas are largely hypothetical and do not get sufficient protection and management attention. The presence of the Forest Department and law-enforcing agencies is either totally absent or very insufficient in most of the CHT. The local people use locally made guns and other kinds of locally made weapons and traps to hunt wildlife for meat. They often use hunting dogs to locate the hunt easily. Nowadays there are many hunters, but very few game animals, leading to overhunting of the game. Yet it is common to find venison, frogs, turtles, lizards and many other wildlife species and wildlife products available for sale in the markets of the CHT. The hill people, especially the children, do not get enough animal protein due to isolation, overall food insecurity and difficulty (or less practice) of fish culture, poultry rearing and cattle rearing. Therefore, hunting is rather essential for the hill people until the substitute is made available.

Conservation of Wildlife

The CHT is a wildlife and biodiversity rich area of world renown, but the status of conservation and management is perhaps one of the poorest of the world. As a consequence, the rate of loss of species and populations is extremely high. The wildlife and their habitats that still exist are basically the remnants of what originally existed. We must act now to reverse the trend and protect whatever is left.

Legal and illegal overexploitation of timber and non-timber forest products must be stopped by creating a multi-stakeholder management committee that will include representatives of different Government departments, national and international non-governmental organizations, representatives of the local government and local people of different communities. Qualified people with relevant background and experience should be transferred or recruited in different entities in the CHT. To some extent, the current trend is to transfer corrupt and incompetent people to the CHT as 'punishment transfer'. Agricultural scientists should study the *jum* cultivation practice and find a way so that the soil fertility can be refreshed in a short period of time and cultivation can be continued. Relocation of people, especially the settlers, from the remaining biodiversity rich areas is necessary for biodiversity conservation. If that is not possible in the present context, people should be motivated and provided assistance for alternative livelihoods so that their dependence on natural resources is reduced. Enrichment plantation can also be initiated to protect the core areas of natural vegetation. Only the indigenous plant species should be selected for plantation so that wildlife can find safe sanctuary.

A strong network of protected areas must be established in the CHT, which will include not only the existing protected areas but also strategically chosen new protected areas and the extension of existing protected areas be made so that all the best quality habitat types are protected. The CHT still has diverse habitats for wildlife, viz. evergreen forests, bamboo groves, bushy and grassy areas, freshwater wetlands and riparian vegetation. We must protect this habitat diversity in order to protect the diverse wildlife that occurs in each of these habitat types.

The hunting of wildlife for consumption and sale is a common practice throughout the CHT. Although the government ordered people to deposit locally-made guns, it is not fully implemented. While all guns should be deposited with the government authorities, traditional hunting by catapult, arrows and bows, and rope-traps (not the commercially made metallic leg-traps) can be allowed outside the protected areas and reserved forests. It is crucial that the farming of cattle, pig, chicken, hare, and fish is expanded and supported, so that the people (especially the kids) get enough protein in the face of reduced hunting. Once these are ensured, a full ban on hunting by all means can be considered.

References

Ahmed, A.T.A., Kabir, S.M.H., Ahmad, M., Ahmed, Z.U., Begum, Z.N.T., Hassan, M.A., and Khondker, M. (Eds.) 2009. *Encyclopedia of Flora and Fauna of Bangladesh, Vol. 27: Mammals.* Asiatic Society of Bangladesh, Dhaka, Bangladesh. 264 pp.

Conservation International 2012. Biodiversity hotspots. <www.biodiversityhotspots.org>. Accessed on 24 February 2012.

IUCN 2011. 2011 IUCN Red List of threatened species. <www.iucnredlist.org>. Accessed on 24 Feb 2012.

IUCN-Bangladesh 2000. *Red Book of Threatened Mammals of Bangladesh.* IUCN-Bangladesh, Dhaka, Bangladesh. 71 pp.

Kabir, S.M.H., Ahmad, M., Ahmed, A.T.A., Rahman, A.K.A., Ahmed, Z.U., Begum, Z.N.T., Hassan, M.A. and Khondker, M. (Eds.) 2009. *Encyclopedia of Flora and Fauna of Bangladesh, Vol. 25: Amphibians and Reptiles.* Asiatic Society of Bangladesh, Dhaka, Bangladesh. 204 pp.

Khan, M.A.R. 1982. *Wildlife of Bangladesh – A Checklist.* University of Dhaka, Dhaka, Bangaldesh. 173 pp.

Khan, M.M.H. 2004. A report on the existence of wild hog deer in Bangladesh. *Bangladesh Journal of Zoology* 32(1): 111-112.

Khan, M.M.H. 2008. *Protected Areas of Bangladesh – A Guide to Wildlife.* Nishorgo Program, Bangladesh Forest Department, Dhaka, Bangladesh. 304 pp.

Khan, R. 2010. *Wildlife of Bangladesh from Amphibia to Mammalia – A Checklist.* Shahitya Prakash, Dhaka, Bangladesh. 128 pp.

Siddiqui, K.U., Islam, M.A., Kabir, S.M.H., Ahmad, M., Ahmed, A.T.A., Rahman, A.K.A., Haque, E.U., Ahmed, Z.U., Begum, Z.N.T., Hassan, M.A., Khondker, M. and Rahman, M.M. (Eds.) 2008. *Encyclopedia of Flora and Fauna of Bangladesh, Vol. 26: Birds.* Asiatic Society of Bangladesh, Dhaka, Bangladesh. 662 pp.

A patch of evergreen forest that still exists in the CHT.   © Monirul Khan

Birds of the Chittagong Hill Tracts

Ronald R. Halder

The entire Chittagong Hill Tracts (CHT) was once brilliantly forested. The natural forest in this unique region is predominantly evergreen and semi-evergreen rainforest, one of the richest biotope known. The diversity of life in these types of forest is staggering. Out of a total bird species count of nearly 700 in the country, more than 450 are found in the Chittagong Hill Tracts.

However, the region is now almost entirely denuded except for a few remaining patches in the extreme north and extreme south. The best-preserved forest patches still remaining are Dhanpata reserved forest and Kasalong reserved forest in Khagrachari district, and Sangu reserved forest in the extreme north of Bandarban district. The main reason for the survival of these patches has been inaccessibility of these places rather than by any conservation effort. One of the most significant contributing factors in the destruction of the hill forest has in fact been accessibility provided by the waterways after the construction of the hydro-electricity dam on the river Karnaphuli. As the artificial Kaptai Lake was created, huge areas became navigable facilitating largescale commercial logging. A pulp and paper mill (Karnaphuli Paper Mill) was established in 1953 in Kaptai in Rangamati district to exploit the huge forest resources just prior to the construction of the dam. Large areas of pristine natural forest were clear felled to make room for teak, rubber and bamboo plantations causing further subsequent destruction of the hill forest.

The destruction of the natural forest has a severe impact on the flora and fauna of this region. Many of Bangladesh's critically endangered bird species are from this area and notably in this list are such iconic birds as the Great Hornbill *Buceros bicornis*, Wreathed Hornbill *Aceros undulatus*, White-winged Duck *Cairina scutulata* and more. As the destruction of this unparalleled forest goes unabated, the hope for the survival of many of the rare bird species in this region hangs precariously in the balance.

There are nearly 470 species of birds recorded from this region up to now. However, it should be borne in mind that this area was mostly inaccessible to naturalists and ornithologists for a very long time due to multiple reasons, notably among them is the armed conflict in the region. Even today, most of the pristine parts of the forest lie unexplored; therefore finding further bird species is nearly impossible. Although it is vitally important to thoroughly explore and catalogue the rich diversity of these beautiful forests, yet the urgency is in its protection; because if we lose these last remaining patches, we shall never know what we have lost forever.

Most of the hill forests are evergreen and semi-evergreen rainforest in nature. These types of forests are best known for their species diversity, and these hills are a testament to that fact. There are birds in these forests that utilize every part of it in some way or another.

We have already lost a number of bird species from this region, but before we discuss that list, we shall focus on the ones we are about to lose, and foremost among this list is the White-winged Duck *Cairina scutulata*. The last known record of a White-winged Duck, a resident of these forests is 1980, no new sightings have been reported since. The reason for the decline of this wonderful duck is most likely habitat loss combined with hunting pressure. This very large duck is a tree cavity nester, meaning, it needs large trees dead or alive with large enough cavity for it to fit in to nest. In addition to large and mature trees, it also needs secluded ponds nearby to raise its young. A gradual decline in the quality of forest and human encroachment of its habitat has possibly pushed it to the point of being exterminated from the country at present.

Another beautiful bird on the brink of disappearing from the country is the Great Hornbill *Buceros bicornis*. Once found extensively throughout the Hill Tracts, it is now limited only to one or two locations. These very large frugivorous birds require mature virgin forests for its survival. They also require large old trees for their nesting activity to be successful. Most of the hill forests have at prescient become unsuitable for their future survival due to chronic forest depletion, and it is a real possibility that we might lose this beautiful bird soon. Pet trade and trophy hunting has also significantly contributed to its decline.

The mountain Hawk Eagle is large bird of prey that calls these forested hills its home. This is a very shy and reclusive bird preferring dense undisturbed hilly forest for its survival. Although there is no previous population estimate to compare to, but by judging their number of occurrence from similar habitat in adjacent countries, their decline in our hill forest is remarkable. This is perhaps the rarest raptor of these hills now. As the quality of the forest declines further, the future survival of this species is seriously in question.

The Green Pigeons of Bangladesh has seen a dramatic decline in their numbers over the years, and some of these species are facing near extinction, and none is more so than the Pin-tailed Green Pigeon. They are possibly the most vulnerable to getting extirpated from Bangladesh among the five Green Pigeon Species. Among the varied reasons affecting their population, habitat loss and natural forest degradation are of primary concern. This species is now the rarest among our Green pigeons. Similarly the Wedge-tailed Green Pigeon is also in serious decline and may also be in a similar situation to the Pin-tailed Green Pigeon.

Possibly once common throughout the hill forest, the Slatey-headed Parakeet must have gone through a serious decline in their numbers. Their population is now highly localized and vulnerable to further decline. Unless research is undertaken, it is not possible to point out the exact cause, but as they are also a species dependent on pristine natural forest, their decline can also be linked to the decline of forest quality, but also at the same time the impact of the pet trade needs to be taken into account.

The Creasted Tree Swift is also a forest dependent species. Their presence in Bangladesh was not known until very recently. They were considered a vagrant species. But a viable resident colony was found in these hills in 2010. As we do not have any previous population estimate, it is not possible to comment on their exact status. But by the look of their flock size and occurrence, this is possibly one species that is about to disappear from our forest soon.

Most of the species facing disappearance from our hill forest have special needs and are therefore most vulnerable to changes in the forest conditions; another species that fits these criteria is the Great Slaty Woodpecker. This is a species that was always limited to these hills where very large trees abounded. As the quality of the forest declined, this species has also declined in number, and has become extremely rare now.

Although the quality of the forest has declined in these hills over the years, yet some areas are still very important for the survival of some

(first row from left) Thick-bill Green Pigeon & Blue-naped Pitta. (second row) Lesser Spotted Eagle & Long-tailed Broadbill. (third row) Jerdon's Baza & Mountain Hawk Eagle. (fourth row) Rufous-necked Laughingthrush & Hooded Pitta.

© Ronald Halder

unique bird species such as the Blue-naped Pitta, Blue Pitta, Oriental Dwarf Kingfisher, Spot-bellied Eagle Owl, Mountain Scops Owl, Jerdon's Baza, Black Baza, Indian Spotted Eagle, Streaked Weaver, Black-breasted Weaver, Gray Peacock Pheasant, Large Scimitar Babbler, Oriental Pied Hornbill, Orange-bellied Leafbird and more. These hills are also an important wintering ground for many of the higher altitude bird species that come down to escape the harsh winter condition of the Himalayas. Some species also arrive from faraway places like Siberia, Inner Mongolia, China and South-east Asia.

The decline of forest dependent species is a global phenomenon now whether it is animals, birds or reptiles. And the reason is always the same, natural forest destruction. Many species are now faced with extinction. One such species of bird that has disappeared from the wild globally that used to call these hill forests its home is the Green Peafowl. It is very unfortunate that this beautiful and iconic bird has disappeared from not only our land but also from all of its former range. We do not know exactly how many species may be in serious trouble within these hill forests. But the outlook for many of the species does not look good. If we are to protect these beautiful animals, we have to protect our forest from commercialization, because that is the greatest cause natural forest destruction. There needs to be a clear demarcation between plantations and natural forest. Not a single patch of natural forest should be destroyed anymore if we are to protect our natural heritage.

It is not only birds that are in serious decline due to forest destruction, large mammals such as elephants, bears, wild dogs, sambar deer and many more are almost at the point of no return. Yet forest destruction goes an unabated. Every year new areas are brought under rubber and teak plantation resulting in further shrinking of the natural forest patches. The process of land sequestration by way of plantation is not only detrimental to the wildlife; it also limits the available amount of cultivable land to the local inhabitants, whereby they are forced to move onto newer areas for crop production. Although the slash and burn method of agriculture that is practised in the hill is not compatible with forest conservation, yet the survival needs of these communities cannot be ignored. Sometimes it is easier to point a finger at this seemingly corrosive agricultural practice than the more insidious method of forest destruction such as land sequestration and establishing commercial plantation. If our wildlife and our forests are to survive in the hills, a balance must be found between our needs and the needs of Mother Nature.

Journey into a Bat-cave

Ronald Halder

As far as I can recollect, the mountains have always fascinated me. I still can vividly recollect my first trip to the mountains of Beluchistan in Pakistan at the age of four. When my uncle called my name aloud to the mountains; the mountain replied back my name. It was sheer magic to me. Then he explained what an echo was.

Even though the hills of Bangladesh are nothing compared to the mighty hills of Beluchistan, they are still equally fascinating. What they lack in grandeur is easily made up for by their beauty and mystery. So every trip I make to our humble hills are equally fascinating today, as it was in my childhood days.

So with great excitement, I and my two other friends, Captain Enam and Rifat head for the hills of Bandarban on the night of 26 April 2007 on a ten-day trek. Our objective is to explore a remote bat cave in a place called Remakri. Captain Enam is a pilot of our national carrier Biman, while Rifat is a graphic designer, and myself a dentist and a nature film maker. Although professionally we have nothing in common, the love for the mountains and the nature united three of us.

After a full night of travel, we reach the small city of Bandarban at six in the morning of 27 April. Our guide and our long-time friend Mong is there to greet us as usual. As we transfer our luggage to the rented four-wheel drive, Mong informs us that we must first go to the house of Lapru-sue, another close friend of ours. A huge Marma breakfast was awaiting us. Lapru-sue is the key organizer of this trip. We were hoping that he would be able to join us, but some other important work excludes him. Although we are disappointed by his inability to accompany us, he assures us that all necessary arrangements have been made for us to travel onward.

After about three hours of Jeep ride we reach the small outpost of Thanchi. This is where the next phase of our journey begins. True to his words, Lapru-sue has arranged a boat and two boatmen for us. Two young Marma men greet us as soon as our jeep comes to a halt near the stony riverbank of Sangu. Although this is my second trip on this trail, the sight of the dugout boat fills my heart with excitement. Now I finally know that this trip is happening.

We make a short trip to the local bazaar and purchase our supplies for the next ten days, then set out immediately. Our boat can barely accommodate our luggage, the two boatmen and us. Yet once the boat starts moving we soon forget the apparent discomfort of our cramped condition. The scenery all around is spectacular. A succession of rolling green hills interspersed with *jum* cultivation and pockets of forest line our way. This is such a complete change of scenery for our eyes accustomed to the chaotic scenes of city life.

Around sun set we reach a small Marma settlement called Tendu. Our lead boatman Aung-shoy immediately sets upon finding a place for us to stay for the night. Tendu is small friendly place and securing a place to bunk for the night is no problem. We find a small eatery run by one of Aung's relative that has a comfortable accommodation on top. From an external view, the structure looks somewhat precarious to me, but as I enter this bamboo structure, I realize the robustness of this hut. Solid logs hold the whole structure from within. To me it feels like that even the severest of storms wouldn't be able to topple it. Except for the external design, almost all homes of indigenous people are built this way.

Each indigenous peoples has its own style of constructing a hut. Therefore it's fairly easy to tell by the look of the structures, what communities the owners are. Even the location of a village can be a clue to the identity of the indigenous community. The Marma would very rarely build their village away from the main river channel, while the Khumi prefers the highest point on the hill, and the Mru the remotest of location. So if you are invited to visit a Khumi village, make sure you have enough strength in your legs, while if a Mru invites you, make sure you not only have strength in your legs but also perseverance.

We start back on our way after a good night of rest at Tendu and stop our boat at a point in the river with a huge rock fall area. The huge rocks that straddles across the river are revered by most of the indigenous communities here, and one huge rock in particular is considered sacred called *Bawng-daw*, meaning king rock. This area is well forested and full of birds and wildlife, so we decide to spend a full day here.

Our two boatmen, Aung-shoy and Kong-la is instructed not to disturb us and do whatever they pleased until late evening when we would decide if to continue further or to spend the night here among these rocks. Our decision to spend time here is rewarded immediately, as we find rare birds nesting in the forest nearby. As instructed, the two boatmen show up around five in the evening, but with a boat full of surprise. I am really amazed by the versatility of this Aung-shoy guy. He has almost single handedly managed to catch eight large carp like fishes from the stony crags in the river. With such good luck, we decide to spend the night here among the rocks in style.

Two large rocks in the river are chosen as our tent pitching ground and we set about making ourselves comfortable for the night. As soon as our tent pitching is complete, Aung-shoy, Kong-la, and Mong set about preparing our dinner.

In my years of travel through the hinterland of our Hill Tracts, I cannot recollect a more memorable meal than this one. I knew that the Marmas are excellent in cooking, but nothing prepared me for the spread that is in front of us tonight. Flame grilled fish marinated with wild herbs, tender fish-eggs cooked in a bamboo joint with fragrant spices collected from the forest, fresh young green edible leaves collected from the forest served with a fiery hot sauce to be eaten as an accompaniment to our meal and a fiery rice wine brew served in small bamboo joints. Our rice is served in washed banana leaves instead of plates. Simply brilliant! I am truly amazed at the culinary expertise of our humble boatman Aung-shoy.

After that fantastic meal, we move back to our respective rock top where our tent are, and lay down to enjoy the cool night breeze and a semi-full moon. Here, the only sounds to interrupt my melancholy thoughts are the twirling of water against the rocks flowing through the mountain stream. Soon I hear a distant call of a Mountain Scopes Owl. Instinctively I pull out my sound recording gear and play back a previous recording of its call into stillness of the night. Almost immediately the bird responds by flying past me in a bullet like speed. Although I have recorded this call on one of my previous trips, but was not able to see the bird, therefore today is my best chance to connect the call with the bird. After a few repetitions of the call, the bird lands at a nearby tree and I am able to look at it with a spot light for the first time while it's calling.

Late into the night as I enter into my tent and lie down, I hear a gun shot nearby. Startled I sit erect in my tent. Captain Enam enquires about the origin of the sound with a slight alarm in his voice. I reason that there must be a hunter hunting nearby. Then after about five more minutes or so, all hell breaks loose. My initial feeling is that a third world war has broken out and we are in the middle of it. Fortunately the noise dies down in half a minuet or so and we realize what has happened. Rock-fall! A sense of unease settles over our tent as we ponder our next course of action. We debate the idea of relocation, but our laziness prevails over our sense of caution and we fall back to sleep.

Next morning we resume our journey at the first light of dawn. Towering peaks and emerald green forest patches line the riverbanks. As the river is very shallow, and the rapids very swift, we must walk most of the way while Aung-shoy and Kong-la drags and pushes the boat up stream. This is quiet a pleasant walk through the sandy riverbanks filled with flowering shrubs and bushes. When we come across a group of young Mru men resting on the riverbanks, we pause our journey to talk and to get to know this group. To our utter disappointment we discover none of them can speak Bangla, so our friend Mong comes to our rescue as an interpreter. In my numerous encounters with Mru men, I have found them to be extremely friendly and hospitable, and this group is no exception; they offer us pan and a strange fruit that I have never seen before. Even our guide Mong has never seen this fruit before. A brilliant glossy dark red fruit about the size of an oversized olive, and as I break it open, viscous dark red fluid fills my hand. They encourage me to put it in my mouth, but I am slightly hesitant, so one of them breaks open a fruit and eats it.

As I put the fruit in my mouth I am pleasantly surprised by the flavor. The pulp is somewhat stringy but sweet. One thing that strikes me about this fruit is its viscous red juice, no wonder the locals call it *Gongoi-see* meaning, bleeding head. This fruit could be used as an excellent food-coloring agent or for making fantastic ice creams. As I enquire, I discover that it's a wild seasonal fruit found in the local jungle only and not cultivated.

I wonder how many more surprises this forest may hold. Depression sets in as I recollect the alarming rate of our hill forest destruction and realize that we will never know what potential food or medicinal plants and fruits we may lose forever.

We reach Remakri at sunset and climb up to a nearby Marma village for the night. Mr. Poosaw, the headman of this village is an old friend of ours and he greets us with an open arm. From my previous trip I recall how he had a special liking for our cans of sardines, so when I present him with a few cans of sardines, his face lights up with an east-west grin exposing his charcoal black teeth from years of smoking piped tobacco. Over dinner we discuss our plan to mount an expedition into the nearby bat cave and request his assistance, which he enthusiastically endorses to.

From this Marma village, the bat cave is about an hour distant and a half of grueling climb and then a terrifying descent into a Khumi village. Our plan is to lug all our gear up to the Khumi village and stay there for the next duration of our trip. As before, Aung-shoy weaves his magic over the Khumi headman and we are granted to stay at his house.

The bat cave is only a short distance away from this village. As soon as we finish our lunch, we head out to scout the cave. We are guided through a narrow tunnel like stream that flows through a dense cover of foliage leading up to the bat cave, and reach the cave entrance in about half an hour of brisk walk. The stench of bat guano is overwhelming and we decide not to enter the cave today, as we are not adequately prepared for the ordeal. But the following morning our sprits are high and our strength back, so with renewed vigor we head for the bat cave. The plan is for me to

A tributary of Sangu River.   © Ronald Halder

enter the cave from the bottom while Emam and Rifat attempts to rappel into the cave through an opening on top.

As I enter this obnoxious hole my attitude mellows a bit. The rocky entrance is beautifully carved by years of water flow, and huge round shaped boulders line my way. Deep pools of water divides section of the interior and I must ford my way through this at some points, while in other points by the help of a bamboo raft made by our men. After I proceed about fifty feet, the spacious hall like ambiance turns into a narrow passageway, I leave my tripod and camera and squeeze myself into a tight crawl space. I can see the passageway leading a fair amount of way, but my enthusiasm to venture further erodes as my eyes starts to sting from the overwhelming gaseous stench of bat droppings. After about hour and a half, I decide to come out, but not before setting some fish hook with baits into the deep pool in search of an enigma. After a brief climb to the top of the hill, I find Enam and Rifat still struggling to secure a safe descent approach into the cave. Finally Enam rappels into the cave and we hear his jubilant exultations.

Today is the last day of our exploration, and our plan is to head back to our boat around mid morning. As I crawl out of my sleeping bag, I hear excitement in Mong's voice. He is urging me to come out quickly and look at my mystery fish. One of the local boys has gone out early in the morning to fetch the fishhook that we laid yesterday and there is good news. This fish is one of the reasons for this expedition here. On my previous trip I had only heard stories about this fish, but this time it's for real.

According to the locals, the fish that we have managed to catch is only a juvenile weighing only about a kilogram and a half, but an adult fish can grow up to be nearly forty kg.

This perhaps is a variety of lungfish that inhabits rocky crevasses of our Hill Rivers and streams. This is my first opportunity to look at this fish, and I am very exited by what I see.

As the sun climbs up, it's time for us to say good-bye and head for home, but at the core of my heart I do not look forward to going back to my city life in Dhaka. But I must return. I envy the simplicity of life among these hills, and hope to return whenever I can. The chance to meet such wonderful people throughout our hills is such a pleasure. What these hills have given me is of great value to me, and I shall cherish them forever as I conclude this trip.

(top) Gongoisee (local name), a colourful and delicious jungle fruit. (bottom) Bounty of nature (mal fish) barbequed by nature lovers.   © Ronald Halder

Brick-Burning in the CHT

In rockless Bangladesh, brick is essential for building concrete houses, roads and highways. However, the manufacture of bricks has serious disadvantages from the ecological viewpoint. It causes deforestation, pollutes environment, and degrades soils. There are legislations restricting the use of wood to burn bricks and setting brick kilns within three km of a forest *inter alia*. But brick kiln owners rampantly violate the law. According to a UNDP report firewood constitutes one-third of the total fuel consumed in the brick making industry in Bangladesh. Brick kilns in the Chittagong Hill Tracts pose numerous environmental concerns.

Burning of Bricks (Control) Act 1989, last amended in 2001, restricts the construction of brick kilns within three kilometers from the boundary of a forest. The recent policy of Department of Environment (DoE) on the construction of brick kilns has rendered any brick kiln construction in all three hill districts—Rangamati, Khagrachhari, and Bandarban—illegal.

According to the estimate of DoE, the number of licensed brick kilns in 2009 was 8,000. According to non-government estimate there were another 3,000 brick kilns around that time (*Prothom Alo* 19/2/2009). This figure matches the estimate of Department of Environmental Science of Bangladesh University of Engineering and Technology (BUET) according to which the number of brick kilns in 2011 was approximately 10,000.

Buddhajyoti Chakma reports on brick kilns in Bandarban and Partha Shankar Saha on Khagrachhari.

Bandarban Situation

There are three legal and 47 illegal brick kilns in Bandarban Hill District. They burn 15,000 tons of wood everyday in violation of the law. These kilns, very close to primary schools, residential areas and reserved forests, also in violation of laws, cause social and environmental concerns. The local people of two Marma villages of Shitbatali in Faitong are under a threat of eviction due to brick kilns.

Fourteen illegal kilns, which are not included in the government list, have been operating in Faitong Union of Lama Upazila. Faitong Union Parishad Chairman and local people informed that 300 acres of forestland have already been cleared of vegetation to supply firewood to the brick kilns. The employers of the forest department admitted that the forest was destroyed for brick kilns. However, they say that the kilns are not the only reasons. There are other reasons too; illegal felling is the foremost among them.

According to the district administration sources, all of 33 government licensed brick kilns are old fashioned except for the *zigzag* processed kilns, which were recently established in seven upazilas of Bandarban. Except for three, the licenses of all other kilns have expired three years ago. Anonymous administration employers informed that these kilns have no chance to renew their licenses because they have marked these as harmful for the environment.

There are 17 illegal brick kilns in Bandarban—one in Naikhongchhari, one in Lama Sadar Union in Lama Upazila, one in Fashiakhali Union, and 14 others in Faitong. These are in addition to 33 government listed brick kilns. The Executive Magistrates sometimes raid the government listed brick kilns and fine the ones violating the law. However, the administration, the police, and the forest department never carried out a raid or interrupted the illegal operation of the brick kilns. An anonymous former member of Faitong Union Parishad said that to get a license or to apply for a license, one has to spend a huge sum of money. The brick-burning season may also elapse by the time one has run from one department to another. However, if one runs the brick kilns completely illegally by having secret deal with the administration, there is no need to go through this expensive and time-consuming process. This is why the owners run their brick kilns illegally. The illegal kiln owners also formed a committee named *Faitong Itbhata Malik Samiti* (Faitong Brick Kiln Owners' Association) to manage the probable "troublemakers" in the administration. The chairman of the association, Fakrul Islam said that they are also members of the Bandarban District Brick Kiln Association and if anyone has any objection relating to the brick kilns, they can contact the chairman of the district association.

Owners' Association: The leaders of the present (2011) ruling party lead the Bandarban District Brick Kiln Owners' Association. It is an open secret that the job of this association is to collect tolls from the owners and use it to manage the administration, police, the forest department officials, political leaders, journalists, and members of security agencies. Although Fakrul Islam admitted collection of tolls, he refused to disclose the amount collected. He said that they needed to give the tolls to run the kilns and the businesses, whether it is legal or not. According to different concerned sources, each brick kiln gives two lacs Taka (approximately 2400 USD). So, some ten million Takas are taken from 50 brick kilns and then the collection is distributed. However, the Chairman of the Brick Kiln Owners Association Mahbubur Rahman denied this allegation. He asserted that the brick kilns are there for the development of the area and everyone contributes money voluntarily.

Impact of Brick Kilns on Local Communities: Though the law restricts setting up of brick kilns within three kilometers of the residential areas, most of the kilns have been built where the ethnic communities live. It was found from spot investigation that half of the kilns have been built very close to the reserved forests, villages, and schools. An example is ABC Bricks situated near the Goungguru Khyang Para Primary School. Chingthui Khyang and Paiklau Khyang, the students of class three of the school described how they faced difficulties in class when the kiln was running. It is very hard to stay in the classroom while the smoke from brick kiln, the loud noise as well as the dust from the heavy trucks come out. Most of the time they get sick and some of them bleed while coughing.

Khyang Student Council sent a petition to the minister for forest and environment on 6 December 2011 with an appeal for shutting down the brick kiln and saving the students from environmental pollution. The chairman of Khyang Student Council, Hlamong Khyang expressed with anger, "Nothing has changed even after informing the administration and the police and petitioning the minister. Khyang leaders also benefit from the brick kiln owners." The *karbari* of Gungguru Para and the chairman of school management committee, Kachingmong Khyang and Khyang leader, Bacha Khyang agreed, "It was wrong not to stop the illegal operation of the brick kilns from the beginning. However, the owners of the brick kilns have already invested a lot into their business. We should consider their loss too."

Two adjacent Marma villages and one government primary school in Shivtali area of Faitong are on the verge of eviction for the same reason. However, an anonymous *karbari* said that they are in severe insecurity and it would not be possible for them to survive. Naliram Tripura, another *karbari*, said that the Marma villagers have become the hostages of the owners, workers of the brick kilns, and the land grabbers. The owners and workers allegedly attacked and harassed them if they said anything against them.

Destruction of Forests and Impacts on Environment: Spot investigation by this writer confirmed the burning of wood in the brick kilns. There are huge stocks of firewood beside the brick kiln chimney. Rafiqul Islam of FBM Bricks informed that they use 400 *maunds* (one *maund* is equal to 40 seer—a seer, an Indian unit of weight, is close to a kg) of firewood everyday if a brick kiln is operating and burning fuel in full force. On an average a brick kiln burns 300 *maunds* fuel wood everyday. According to this counting, 15,000 *maunds* daily, 450,000 *maunds* a month and 2,250,000 *maunds* firewood are burnt in one season [of five months] in 50 legal and illegal brick kilns in Bandarban.

An anonymous *headman* informed that everyday thousands of *maunds* of wood is cut from reserved forest of Fashiyakhali Union and sent to the kilns. The owners of the brick kilns have a secret deal with the administration and the police officers; they do not care anyone else. The Union Parishad chairman of Lama Faitong, Shamsul Islam informed, "In spite of having complaints about the illegal kilns, the administration does not take any step. Millions of takas have been spent on 320 acres of land of the reserved forest for plantation in the financial year 1986-87. Use of massive quantities of wood in brick kilns has led to destruction of forest in the reserved and unclassed state forests."

Mir Kashem, Range Officer of Faitong Doluchhari Range agreed that all but 30 acres of 320-acre reserved forest have been destroyed. "It is not just unstoppable forest marauders, *jum* cultivators also slash trees," said Kashem. "We cannot effectively work for the lack of man power and logistical support such as vehicles and fire arms."

It is not just forest that is damaged. Angry about harvest of soil from the hills and agricultural land Shamsul Alam complained, "The hillocks are leveled and cropland is mined for soil, essential for brick making."

Government Actions: The administration and the Department of Environment of Chittagong sometimes raid the illegal brick kilns when media reports come to their attention. The authorities impose some fines on the owners and stop stockpiles of wood. Then business of brick burning in violation of law resumes as normal. The raids are just eyewash. The government passed an order in 2009 to close the brick kilns not following *zigzag* method in order to protect the hills and forests. Nur Hossain, owner of the Fashiyakhali HDB brick kiln and Fakrul Islam, Chairman of the brick kiln committee of Faitong frankly admitted that they do not have legal papers for their old-fashioned brick kilns. According to a 2010 government order brick kilns, not converted to *zigzag*, method must be stopped. Renewal of license for the old-fashioned brick kiln was also stopped.

Asaduzzaman, Alikadam Upazila Nirbahi Officer said, "We are aware of the official order to close down all old-fashioned brick kilns nationwide. Any such kiln operation is illegal."

The Deputy Commissioner (DC) of Bandarban, however, was found soft about the old-fashioned brick kilns. "If the illegal kilns are closed down, the infrastructure development in remote and un-developed Bandarban would be impeded. Considering this, we are relaxing the law relating to the brick kilns," said the DC who was found aware of operation of illegal brick kilns in Faitong Union of Lama. The DC reportedly did not receive any complaint about illegal brick kilns.

Most of the owners of 50 legal and illegal kilns are linked to political parties, especially to the incumbents. An allegation about the leaders of brick kiln owners' association is that the leadership changes with the change in the party in power. The kilns owned by party cronies enjoy leverage.

Harun-Ur Rashid, the Lama Divisional Forest Officer, and Mir Kashem, Range Officer, Faitong Doluchari claimed that it is not the duty of the Forest Department to give licenses or control the brick kilns. They take rapid action when they get complaints about firewood use in brick kilns. They believe it is difficult to stop the unscrupulous forest bandits. However, the Forest Department was considering giving away the 30 acres of the remaining forest in Faitong now under the social forestry program.

Khagrachhari Situation

Two brick kilns are sure to draw the attention of the visitors traveling along the road from Khagrachhari sadar to Panchhari upazila. The two brick kilns—BAM Bricks and Satata—face each other in Nalkata village under Panchhari upazila. Of 32 brick kilns in Khagrachhari, only two have government licenses (renewable every year), says a leader of the Khagrachhari Brick Kiln Owners' Association. The BAM Bricks is one of the two government approved brick kilns in the district. However, this kiln burns firewood in violation of the law.

"I supply firewood [to BAM Bricks] during the whole brick making season," said Osman Ali while his supply of firewood was being unloaded from a rickshaw van. A young man, claiming himself as the agent of the brick kiln owner, was weighing the firewood brought by Osman. "This brick kiln burns fire wood," was the candid confession of the agent. He, however, got angry with us as we attempted to take photos. He left us and returned after a while with another young man who began to question us—who we were, what we wanted to do, and in particular whether we were journalists. The agents then made a call over the cell phone and attempted to force us to stay there. We tried to cool them down and managed to escape.

Our experience of visiting BAM Brick was an unpleasant one. BAM has a license, and yet the men associated with the kiln turned angry. Our fault was we saw the firewood burned at the kilns. They probably feared legal actions from the authorities if we reported the incident in the newspapers.

According to section 5 of the Brick Kiln (regulation) Act 1989, "No person shall use fuel to burn brick". Its section 2 (ka) defines 'fuel' as 'any kind of plant fuel except for dried bamboo stumps.'

The law stipulates that only a Deputy Commissioner (DC) of a district can issue a license for setting up brick kilns. The DC himself and his/her representative(s) can inspect any brick kiln at any time. The district authority can seal a brick kiln if it is found burning wood, says the law.

The Khagrachhari district administration does not know the number of the brick kilns in the district, admits Khalilur Rahman, the additional deputy commissioner (ADC). But he is aware the brick kilns burn firewood. He raided several brickfields at different times and seized huge quantities of firewood and bricks. "We will take immediate steps to stop burning of firewood in the brick kilns," said the ADC.

Though the district administration doesn't have any information on brick kilns, the Upazila Nirbahi Officer (UNO) of Khagrachhari Sadar upazila, Rahat Hossain keeps record of brick kilns in his upazila. Hossain says that every brick kiln has a license but none has been renewed after they were issued. "Considering the demand for bricks in the area no steps have been taken," says Hossain.

It is obligatory for any brick kiln to use a 120 feet chimney, yet brick kilns in Ramgarh, Mahalchhari, and Dighinal upazilas still use drum chimney as reported in newspapers and told by the brick kiln owners' sources. Drum chimney is banned because it releases more pollutants than others. This writer found some brick kilns—KBM in Baillachhari, BBS in Wachu and NMB in Kamini member para in Matiranga upazila—using drum chimney and burning wood. All of these kilns had wood, mostly fruit and timber trees including *gamari*, piled up in their premises.

The people associated with the brick kilns reported that while fruit trees come from homestead gardens, *gamari* is stolen from forest department's plantation.

Kamal Saodagar, owner of KBM in Bailachhari confessed that he does not have any license for his kiln. But he still manages the administration not to take note of his illegal operation.

Seen from a hill in Alutila area the landscape on the western outskirts of Khagrachhari town with a number of brick kilns.   © Philip Gain

Invasion of Tobacco in the CHT

Tobacco cultivation started in the CHT around 1990 and has expanded quite rapidly. It has taken most of precious valley land in Bandarban in particular. It is also creeping into Khagrachhari. Buddyojyoti Chakma reports on the invasion of tobacco into Bandarban and Partha Shankar Saha on Khagrachhari.

Tobacco in Bandarban

Tobacco cultivation is on sharp rise in Bandarban Hill District. In the absence of government and non-government efforts to contain it, the farmers, willingly or unwillingly, are getting into tobacco cultivation. Many agricultural experts agreed that in 2011, most of 75% of land suitable for spring crops came under tobacco cultivation, although according to the Department of Agriculture, tobacco was cultivated in less than seven per cent of the land. It is no new knowledge that tobacco cultivation leads to a decrease in soil fertility, along with air and water pollution. In addition, it is feared that much more of the forests are now under the threat of destruction. All of these combined can lead to irreversible ecological disasters, public health damages, and human misery.

According to the Department of Agriculture Extension (DAE), approximately 950,000 acres of land in seven upazilas of the district 134,412.97 acres (54,395 ha) are suitable for farming crops. According to the officials of Department of Agriculture (Krishi Bibhag), in 2011, while the temporary fallow land occupied 17,037.50 acres (6,894.8 ha), *boro* (winter rice) was cultivated on 16,225 acres (6,566 ha), *robi* (winter) crops on 16,625 acres (6,727.9 ha), and tobacco on 8,455 acres (3,421.6 ha) of the cultivatable land. Even though the locals and the tobacco cultivators claimed tobacco farming is on the rise every year, officials of the agriculture asserted that its cultivation has decreased by 1,737.15 acres (703 ha) in the 2011 compared to the past year. Statistics produced by Department of Agriculture Extension suggest *boro, robi*, tobacco and fallow land covered 58,342 acres (23,610 ha) of the cultivable land. The officials of DAE failed to give any information about is cultivated in 77,645 acres (31,421.8) of cultivable land during the *boro* or *robi* season.

Nurul Islam and Hossain Ahmed have been cultivating tobacco for 12 years in Choto Bamu Shapmara Jhiri of Lama. They claim that tobacco cultivation increased significantly in 2010. They also informed that nothing but tobacco had been cultivated in a 30 km span area from Chokoria of Cox's Bazaar, to Chhoto Bamu and Boro Bamu of Lama, ending near the Chimbuk hills.

One of DAE officials, seeking anonymity, confirmed that almost three-fourths of the cultivable land was being used for tobacco cultivation. This means in addition to 8,455 acres of land according to official disclosure (in 2011) tobacco was cultivated in most of 77,645 acres (31,421.8 ha) of land outside the official account of 2011. Field investigation confirms such widespread tobacco cultivation.

Tobacco companies are directly involved in farming tobacco in the district. For pure business gains and purposes, these farming companies always refrain from divulging information on how much tobacco is cultivated, their budget, and the amount they loan out to local farmers.

Inside Intel: Shamim Jahidi, head of public relations of British American Tobacco (BAT) informed that Dhaka Tobacco, Abul Khayer Company Ltd., Azizuddin Company Ltd., Akij Company Ltd., Nasir Tobacco (Nasir Gold), Rahim & Karimuddin Bhorosha Tobacco, Liyakot Tobacco, New Age Tobacco, and Ali Tobacco along with BAT were amongst the many companies cultivating tobacco in the seven upazilas of Bandarban Hill District. Jahidi refrained from disclosing the amount of land BAT and other companies used for tobacco farming. A high official of BAT, on condition of anonymity, informed that the company was cultivating as much as 1,500 acres of land of 2000 farmers, and the company's share was 36% of the total cultivation. Field officers of other companies, on condition of anonymity informed that they had 10% more cultivators than the registered ones.

According to their information, in the Sangu river valley (Bandarban, Ruma, Roangchhari, and Thanchi upazila) 934 farmers of BAT, 700 of Dhaka Tobacco, and 174 of Abul Khayer Tobacco were cultivating tobacco. No information was found about land under tobacco cultivation, its production, and the number of farmers involved in tobacco cultivation in Lama, Alikadam, how much tobacco is grown, and how many farmers were associated with tobacco cultivating.

Sonali Bank's Data on BAT Cultivators: All transactions of tobacco cultivation for BAT are made through Sonali Bank annually. The manager Md. Rafikul Talukdar of the Bandarban Branch says that loans were provided directly to the companies, including BAT, till 2010. Orders from Bangladesh Bank have stopped these direct loan facilities for tobacco cultivation since 2011. Even then, big companies such as BAT still have transactions pending with the Sonali Bank along with savings accounts being opened for each farmer. This situation puts the system through a loophole that compels the bank to still hand out loans, despite Bangladesh Bank orders. The families registered with the bank's local branch in 2010 were 2,560, which increased to 3,639 in 2011. The table below shows the number of farmers availing loans in 2010 and 2011:

Upazilas	Cultivator family		Amount of loan (Taka)		Creditor	
	2010	2011	2010	2011	Sonali Bank	Company
Bandarban, Thanchi, Ruma, Roangchhari	921	934	12,100,000	20,000,000	2010	2011
Alikadam	488	400	7,500,000	7,486,000	2010	2011
Lama	504	1,105	9,000,000	40,000,000	2010	2011
Naikhongchhari	647	1,207	7,613,000		2010	2011

Besides in Shangu valley Dhaka Tobacco lent Taka 11,000,000 and Abul Khair Tobacco Taka 1,500,000. ™

Local people and tobacco farmers all sing in the same tune as to how the tobacco companies along with DAE never reveal the actual amount of land used or even the actual amount of tobacco produced. The lands in Lama, Alikadam, Naikhongchhari, Thanchi, and Ruma are all blue–covered with tobacco plants; only the wet marshes are left alone.

Md. Nurul Islam of Shapmara Jhiri, Lama, says that the companies refuse to register farmers who do not have at least one acre of land in their names. Islam and Mojong Tripura, head of the Bodhujhiri Para of the Gajaria Union in Lama, say registered farmers farm on lands ranging from one to 10 acres each. Tripura himself cultivated tobacco in 4.5 acres in 2010.

Tobacco: Why and How Long? Tobacco cultivation was a traditional practice for the ethnic communities of Bandarban. They used to farm tobacco along with other spring crops such as eggplant, potato, cabbage, tomato, etc. along the strips of sandy land rising out of the beds of the Shangu and Matamuhuri rivers. Tobacco was treated as any other crop that they used to raise with no business motive. Chingthui Marma of Tindu Bazar, Thanchi, and Uhainu Marma of Remakkry say the locals used to produce tobacco in the area even before the companies started to arrive some 15 years back. The surplus tobacco at that time used to be taken to the markets just as any other spring crop.

The companies barged into the scene with incentives such as market security, agricultural inputs and facilities, and loan assistance to attract the locals into farming tobacco as a trade. Eventually these farmers had to forget about actual crops because of the lucrative business tobacco brought in. According to DAE, the companies started to give loans to the farmers since 1994. They even stood as guarantors for farmers taking loans from banks. This assured the lending institutions to hand out the cash. In the short run, the locals found tobacco cultivation profitable on their fertile lands and the availability of firewood made the cultivation lucrative.

A farmer having an acre or more land can register with the companies. Mushfiqur Rahman, Bandarban BAT Manager, and Paritosh Nag, Manager Dhaka Tobacco, informed that the loans vary from Tk. 5,000 to Tk. 30,000 depending on how much land a farmer devotes for tobacco cultivation. The farmers are also provided with seven bags of different fertilizers, including Urea, TSP, and SOP, for each acre of land they own.

What attracts the farmers most for tobacco cultivation is market assurance. Mongkoai Marma Roangchhari and Mongbachi Marma from Thanchi say that they prefer growing something that is sold from their homes, rather than crops that can perish if not sold. The companies are always there to buy off tobacco from them for handsome prices. A farmer growing vegetables and other crops has to carry the produce to the market, wait, and crops not sold may rot. But tobacco has no fear of rotting and can be safely stored for weeks. Such market stability is always desired, and hence the farmers resorted to tobacco farming.

Nasima Baknu Shyamoli, head of Tobacco Control Project IPSA (an NGO), says the farmers cultivate tobacco despite knowing it is harmful for the soil. But this is overlooked because the companies end up giving the farmers huge amounts of money at a time to cover costs for the whole season. The poor farmers value the availability of cash over damages that tobacco does to the soil. Farmers such as Sathoyaiching Marma and Kalachan Chakma from Bolipara of Thanchi say that acquiring a loan from anyone is easy as long as one is affiliated with the tobacco industry. This helps them in fulfilling their daily needs as even shopkeepers allow credit.

Mongkoyaipru Marma from Bandarban Sadar and Al Hossain from Alikadam say they would rather cultivate vegetables only if they were as lucrative as tobacco. Many other farmers from Thanchi, Ruma, and Lama are of the same opinion as there is no market assurance for other crops. On top of that, the agricultural officers from tobacco companies provide door-to-door technical support for tobacco cultivation, whereas crop farmers do not get the same services from the government agricultural offices. And as they lack direct communications with wholesalers they are cheated. Even though the government has stopped loan facilities for tobacco cultivation, there are no incentives for vegetable farmers. Tobacco already being a lucrative option stops farmers from cultivating other crops.

Tobacco cultivation is labor and so it is located in areas where many people live under the poverty line. Businessmen such as Abul Boshor from Thanchi lose out, as they are unable to hire workers as they already work for the tobacco companies. Boshor has a hard time recruiting labor even after offering higher wages as the tobacco companies hire them with wages in advance. Labor in this sector is in so much a demand that even children are put to work, sacrificing their education. Work for Better Bangladesh (WBB) Trust, in a survey on tobacco cultivation and the possibility of alternative crops in a state of poverty, found out that tobacco farming demands a lot of labor. The tobacco farmers engage their children in drying tobacco during the harvest season. This bars the children from going to school, which is very harmful for them. Such harm is not taken into account.

Tobacco Cultivation: Who Benefit, Who Lose? Mojong Tripura of Badujhiri Para, Lama, spent Tk.180,000 (USD2,168)) on his four acres of land for tobacco cultivation. He expected a return of at least half million Taka (USD6,000). Similarly, Balipara Union Council member Sathowaiching Marma says farmers earn Taka 16-17 lacs (USD19,200 to USD20,400) every season from farming tobacco in contrast to the previous Taka five-six lacs (USD6,000 to USD7,200) from cultivation of rice. Managers of BAT and Dhaka Tobacco both say that farmers find tobacco cultivation more profitable and thus are more inclined to the trade.

Agriculturists and environmentalists, having acknowledged the economic profit, still point out how harmful tobacco farming can be. They claim there is nothing to be gained, but everything to be lost in the future with this industry. Altaf Hossain, principal scientific officer of the Cotton Development Board and former agriculture officer of Bandarban Sadar says tobacco cultivation leads to damage of soil, air, water, health, and ecology. The soil loses its fertility with the use of fertilizers as they end up killing many microbiological creatures that constitute the health of the soil. The fertilizers and pesticides, when washed off the land, flow into the surrounding water bodies causing severe water pollution. Shafiur Rahman, Lama Upazila Health and Family Planning Officer says pollution caused by tobacco farming can lead to an outspread of pneumonia and asthma. While environmentalist Zubayer Al Arman says that the loss of the surrounding woods being a source for firewood needed for drying tobacco is irreplaceable no matter how much money is offered for damages.

Environmental Hazards Due to Processing of Tobacco: Processing tobacco includes a very intricate method by which the leaves are conserved by applying heat to them. A special type of oven, called *tandur*, is used to serve this purpose. A *tandur* is assigned for as much as two acres of land. Fomaching Marma and Nasima Begum from Sainga of the Bandarban Sadar Union built one such oven using as many as 120 trees of different sizes and around 500 bamboo. The *tandur* uses up more than 600 *maunds* (one maund is 40 kilograms) of firewood every season. Karbari Majong Tripura from Bodurjhiri para under Lama made two *tandurs* for 4.5 acres of land. A BAT official seeking anonymity has also confirmed that on an average two acres of land need one oven. According to him there are 17,000 of these *tandurs* in operation spanning across all the seven upazilas in Bandarban. This in Bandarban Hill District 408,000 metric tons of firewood is burned every season for drying tobacco! This huge amount of firewood comes from the surrounding forests in the area. A *tandur* lasts for several years.

Siyongong Mro, a researcher on land management in the CHT fears disaster in the near future if preventive actions are not taken immediately to save the forests. Only 3.1% of the land in the CHT is suitable for crop cultivation, as most of it is steep in nature. Thusly, a significant percentage of the local indigenous people are dependent on the forests; and *jum*, their traditional agricultural practice can only be performed on the forestland. At this rate of forest depletion and loss of soil fertility, the locals will soon be stripped off their income sources.

Environmental Crisis: The river Matamuhuri's ecology is changing due to all the pollutants being introduced to it along with waste tobacco plants and leaves. The fresh water shrimps are in danger of being eradicated due to the washed away residue of chemical fertilizers and pesticides, and people bathing in the river are contracting skin irritations. Tobacco cultivation is now posing serious threat to the people and the biodiversity of the area.

Litigation against Tobacco: A journalist and an employee of an office filed a lawsuit in September 2010 in Bandarban Joint District Judge's Court against the tobacco cultivation of BAT, DAT and Abul Khair Leaf Company (AKT). The three companies appealed against the suit. The court, in an order, restricted the tobacco cultivation of these three companies to 1,000 acres of land. The court also ordered that no forest resources be used for processing tobacco. Shamim Jahidi, Head of BAT public relations department informed that the order given by the District Judge's Court has been postponed in the higher court. This has given a kind of legitimacy to tobacco cultivation in Bandarban and nothing could be done about tobacco cultivation until the case was settled.

The deputy director of DAE, Bandarban, Shamsuddin Ahmad, claimed that the expansion of tobacco cultivation has subsided due to the lawsuit, the increase in the price of paddy, and a cut in government supplies of fertilizers. Besides, he claimed, the Forest Department halted tobacco cultivation in the Matamuhuri Reserved Forest areas. Lastly, Ahmad claimed that even without specific government direction, locals were being made aware of the cons associated with tobacco cultivation and the farmers encouraged to get back to other crop cultivation.

Tobacco in Khagrachhari

River Myani in Myani Valley flows on the east of Dighinala upazila town of Khagrachhari. Both sides of the river have strips of arable plains land, very precious in between hills of the CHT where hardly two per cent land is arable. It was not long ago that farmers grew vegetables and paddy on these strips of land. But now, tobacco covers this land for eight months of the year.

"I used to cultivate potato with vegetables. But when tobacco came we gave up cultivating vegetables. Tobacco is more profitable than other crops. It brings cash. We can sell it for sure. So, we all have a tendency to cultivate tobacco," said Dayaranjan Chakma, a farmer who lives on the bank of Myani river.

Dayaranjan, owner of more than one acre of land, has been cultivating tobacco for two years now. The big loss in potato production forced Dayaranjan to cultivate tobacco. Two years back he grew potatoes abundantly but the price was very low. There was not a single cold storage in the whole upazila. "To make up the loss I began tobacco cultivation," said Dayranjan. "I intend to continue tobacco cultivation."

In 2011, Dayaranjan produced 930 kgs of tobacco on his one-acre land. He sold 800 kgs for Taka 35,000 to the British American Tobacco (BAT). The BAT did not buy the remaining 130 kgs because of its low quality.

According to Khagrachhari district agriculture extension office, during 2010-2011, tobacco was cultivated in 951 ha of land in Khagrachhari district. During this time the arable land under tobacco cultivation in Dighinala upazila was 680 ha, which meant that 71% of tobacco of the entire district was cultivated in this upazila. Next to Dighinala, Matiranga upazila grew tobacco on 218 ha in 2010-2011. The other six upazilas of the district grew little of tobacco.

Tobacco seeds are sown in the month of October. The crop is harvested

Tobacco in a Chak-inhabited area in Naikhongchhari upazila. Once the Chaks used this land along a stream for grownig vegetables and other crops.   © Philip Gain

in April. "If the seeds are good, 30 decimals of land produce 600 kgs of tobacco," says Amanul Islam, a farmer of Kabakhali Muslim Para. The tobacco companies usually supply essential inputs such as seeds, polythene, fertilizers, and pesticides used in the field. The company's trained people advice the farmers about farming.

The risk is less in tobacco production, farmers say. However, they stay fearful of hailstorm and fire.

After harvest, the most important thing is baking of the green leaves. A square shaped house with mud walls, locally known as *tandur* is used to bake tobacco leaves. There is an oven at a corner of *tandur*. Bundled Tobacco leaves tied are hung from hard sticks for baking. After baking 8 to 9 kgs of green leaves are reduced to one kilogram of dried tobacco. Near about six hundred bundles can be dried at a time at in *tandur*. It takes 72 hours to burn such an amount of tobacco.

It is company propaganda that *dhanicha* (a jute-like fibrous plant) is used for baking green leaves. Paris Chakma, who has a *tandur* near the river Myani, informed that he has never used *dhanicha* to bake tobacco at his *tandur*.

On the one hand tobacco has taken the land of vegetables, and on the other, the local environment is affected due to the use of chemical fertilizers and pesticides. Poisonous element released by the tobacco leaves is also harmful to arable land.

Mohan Kumar Ghosh, agriculture officer of Dighinala upazila informed that due to tobacco cultivation, vegetable cultivation has decreased by 300 ha compared to the year past. "Dighinala used to be a 'storehouse of vegetables'. Fulfilling the local demand the surplus produce was sent to other areas. Now, we import vegetables," said Mr. Ghosh. He thinks that it is possible to bring back the local farmers from tobacco to vegetable if a cold storage is set up.

The district agriculture extension officer Bhabotosh Sarkar concurs with Mr. Ghosh. "I can't give you exact statistics but there is no doubt that tobacco has a severe impact on vegetable production," said Mr. Sarkar. "In the tobacco field OSP fertilizer is indiscriminately used. It increases acidity in soil. Besides, excess Urea and Ammonia sulphate are used to make tobacco leaves big. It also deteriorates the soil quality. The tobacco residue that falls on the ground is harmful for the soil. Tobacco plants extract more nutrients from soil than vegetables. Its roots are also deeper".

Environmental and Health Effects

Environmental Effects: It is not known how many *tandurs* are there in Khagrachhari. According to government statistics, tobacco was cultivated in 951 ha of land during 2010-2011. Three tones of firewood are required to burn tobacco grown in one ha of land which means approximately 2853 tons of firewood was burnt to bake tobacco leaves in Khagrachhari. Forest trees along with the trees from homestead gardens are used to bake tobacco leaves, claim owners of *tandurs* and the farmers. An officer of Khagrachhari forest department, seeking anonymity, said, "It is alarming that wood is used to dry up tobacco leaves. We shared our concern with the higher authority several times. But no measures have been taken."

Health Hazards: Doyaranjan Chakma is least bothered about the negative impacts of tobacco cultivation. He has to endure the sharp smell of tobacco leaf while working in the field. He very often catches a cold resulting in frequent sneezing. But the health hazards are immaterial to him.

Rahima Khatun (38) works at a *tandur*. Her job is to keep bundled tobacco leaves into the *tandur* and bring these back when the baking is complete. "There is a sharp smell after tobacco is baked. Initially I became sick by the smell but now I have become used to the environment," said Rahima.

Mostly women and children are engaged in collecting tobacco leaves and in their processing. Dr. Shahid Talukdar, health and family planning officer of Khagrachhari sadar upazila said that the people engaged in the tobacco processing will have to suffer as the smokers do. The smoke released from the tobacco oven is sharper than that of the cigarette smoke. This can cause lung cancer and increase the risk of heart diseases. The pregnant women are in prime risk".

A group of researchers of Department of Farm Power and Machinery of Mymensingh Agriculture University have blamed tobacco and *jum* cultivation as the two major causes of environmental degradation in the CHT. The researchers, led by Dr. B K Bala, conducted a study from November 2008 to May 2010 in the CHT. Tobacco cultivation has been marked as one of the biggest threats to food security in the CHT. The study funded by Food and Agriculture Organization (FAO), has recommended that strict restrictions be immediately imposed on tobacco cultivation in the hilly areas.

A Chak stands on land that was his homestead years back. Now the outsiders have taken possession of his land and are cultivating tobacco. © Philip Gain

Timber and Furniture Trade

Wood brought out of the CHT in the form of timber and furniture, legally and illegally, is a major environmental concern. The reserved forests were created during the colonial times for exploitation of forest produce. Timber extracted from the CHT "was used as sleepers for the railways in the plains (railways were never constructed in the Chittagong hills) and for shipbuilding in the port of Chittagong" (Schendel, Mey, and Dewan 2000: 131). This is evidence how forest were exploited for industrial use. During the Pakistan time the Forest Industries Development Corporation (FIDC) was created (by and ordinance of 1959) for clear felling from the reserved forests and to facilitate plantations that the Forest Department carried out. FIDC was renamed as Bangladesh Forest Industries Development Corporation (BFIDC) after independence and the logging operation at the behest of the state continued. However, the forest resources gradually became so exhausted that the official extraction of timber was completely suspended from early 2000. But the hunger for wood from the CHT [particularly teak for furniture] gives impetus for cultivating timber trees in personal gardens or *jote*. In addition to the harvest of wood from personal gardens with permits, the last trees of the reserved forests are also reportedly illegally felled. There are laws, regulations, and watchdogs to restrict illegal felling and smuggling of wood from the CHT. But it still happens everyday and in massive quantities. Most of the furniture factories, show rooms, and brick kilns seen in the CHT are illegal. Buddhyajyoti Chakma report on trade in timber and furniture in Bandarban and Syeda Nusrat Haque and Supryio Chakma with Philip Gain report on the Rangamati situation.

Bandarban: Natural Forest Exhausted under Cover of Timber Trade

Wood going out of four of seven upazilas in Bandarban Hill District every month is estimated at 200,000 cubic feet. Wood is taken out both legally and illegally. The quantities of timber going out of Lama, Alikadam, and

A lakeside timber depot in
Rangamati town. © Philip Gain

Naikhongchhari upazilas is not known.

Sources in the Forest Department and those involved in wood trading informed that individual owners of timber gardens in Bandarban division, Bandarban Pulpwood Division, and Lama Forest Division get *jote permit* to sell their wood. *Jote permit* is permission from the authorities granted to the individual garden owners for harvest, sale and transport of cultivated wood on individual land and land leased from the DC's office. The sale and transfer of wood with *jote permit* is legal. On the other hand, timber extraction, harvest and transport are prohibited from the natural forest. However, half of the wood harvested with jote permit in hand allegedly comes from natural forests. A wood trader, on condition of anonymity, confessed that this mechanism of legalizing the illegal extraction from natural forest is an open secret.

The 'royalty permit' for wood extraction and trade remains suspended from the 1990s in order to control wood smuggling. After suspension of royalty permit, *jote* permit remains to be the only legal mechanism for timber extraction and its trade. The *jote* permit holders do not need to pay any tax to the government.

Timber trade—legal and illegal! While *jote* permit sets the legal framework for the timber trade, the manufacture and trade of furniture in most cases are illegal. There are allegations of abuse of *jote* permits that allow direct timber trade. A huge quantity of timber slips out of the CHT under the guise of *jote* permits.

The Bandarban Divisional Forest Officer (DFO) Towhidul Bari Khan informed in May 2012 that 35,396.59 cubic feet of wood had been transported by 65 trucks from Bandarban Forest Division, Bandarban Pulpwood Forest Division and Kaptai Pulpwood Forest Division. In the dry season the number of trucks loaded with wood exceeds one hundred a month, informed the DFO. The Bandarban Timber Merchant Association's office secretary Sirajul Islam puts the figure even higher—on average 125 trucks transport timber out of the CHT every month. The above statistics imply that 65,000 cubic feet of timber are legally transported from four upazilas of the Sangu valley by trucks with a capacity of 520 cubic feet in each.

The illegal timber trade and smuggling is widely speculated. The timber merchants who prefer staying anonymous said that equal quantities of timber and wood legally traded are illegally transported through Tonkaboti, Shoroi, Baishari, Shuwalok, Kanaijopara Mongjoy, Dhopachhari, and many other places in Bandarban. This timber comes mostly from already exhausted natural forests.

The furniture business in Bandarban is nothing but another way of the timber business in disguise. The president of furniture shop owners association Syed Ahmad Talukdar informed that there are 200 furniture shops in Bandarban city of which 173 are association members. However, 56 have trade licenses and 60 shop owners have applied for license. Twenty-seven shops that are not members of the association along with 57 members of the association (84 in total) have not yet applied for license. The president of the association has no idea about the number of the furniture shops outside the town.

The treasurer of furniture shop owners' association Md. Alam informed that the DC's office issues license for shops on the basis of a report from the Forest Department. The fee is Tk.200 (USD 2.8) paid through treasury *challan*. This license allows only manufacture and display in the showrooms. According to the Chittagong Hill Tracts Forest Transit Rules, 1973 one has to secure a separate permit for transportation of furniture out of the CHT. This transit rules makes it an obligation for the transporters to clearly mention the name of the *jote* landowner and *jote* permit against which the timber has been procured and how the forest produces have been manufactured.

Syed Ahmad Talukdar and Mohammed Alam confessed that they use wood from sources other than the *jote* permit. One such source is the forest dependent poor local people who bring small pieces of wood on shoulders. Another source is sawmills—shop owners who procure wood without any permission.

Transfer of furniture manufactured with illegal timber: The DFO of Bandarban explained three situations for issuing transfer permits. First, an employee or official transferred from Bandarban to other places is permitted to take his or her furniture. Second, furniture can be transported as wedding gifts. Third, a family migrating from the hilly region to the plains land can take its furniture. In these three situations, the legality of the wood is not questioned. The leaders of furniture shop owners' association refused to disclose the number of trucks that transfer furniture outside Bandarban. However, information from different sources including laborers working in the furniture shops suggests that every month some 400 truckloads of furniture are transferred from 200 furniture shops of

Bandarban. Each of these trucks can carry furniture of 125 cubic feet of wood, which means roughly 50,000 cubic feet of wood are transferred each month in the form of furniture.

The FD officials are unwilling to talk officially about the furniture shops and the furniture business. Unofficially they say if the forest laws are strictly followed in issuing license no one would be willing to get into the furniture business.

Saw Mills: The presence of a large number of sawmills in the Chittagong Hill Tracts demonstrates a deep ecological concern in the Chittagong Hill Tracts (CHT). The names of 28 sawmills were reported from Bandarban Sadar, Ruma, Lama, and Roangchhari. Another five sawmills in Alikodom and seven in Naikhongchhari were reported, but their names could not be found. Bangalis own most of these sawmills.

Timber and Furniture in Rangamati Hill District: An Ecological Audit

Timber harvested and traded from the privately owned gardens and the reserved forests is an ecological concern in Rangamati Hill District that witnessed inundation of huge tracts of natural forests due to the creation of the artificial Kaptai in the 1960s. The temporary military camps and the Bangalis (nearly half million) settled in the CHT also destroyed much of the forests. Yet, a large percentage of people of the district depend on timber for their livelihood. The furniture manufacture and its trade controlled largely by the Bangalis also go non-stop day and night. All these have serious ecological consequences.

The Rangamati circle of the Forest Division comprises of six divisions— (I) Chittagong HT North Division, (ii) Chittagong HT South Division, (iii) Unclassed State Forest Rangamati Forest Divison, (iv) Jhum Control Forest Division, (v) Pulpwood Kaptai Forest Division, and Khagrachhari Forest Division. A divisional forest officer (DFO) who heads a division has the authority to allow extraction of timber up to 55,000 cft per month, which means that 275,000 cft of timber can be legally extracted each month in the Rangamati Hill District. Annaya Shadhon Chakma, general secretary of *Upojatiyo Kath Baboshaye o Jote Malik Samiti* (Tribal Timber Trade and Jote Welfare Association) believes that illegal extraction of timber amounts to 30% on top of what is legally permitted. These illegal woods either go directly to the sawmills or are used to produce furniture.

The additional deputy commissioner (ADC), Mr. Shafiqul Islam, in response to a query about the writ that was filed against the DC office by *Upajatiyo Kath Babshayee o Jote Malik Kalyan Samiti* in 2012 for holding spot inspection by the joint spot inspection team under the Upazila Nirbahi Officer in *jote* permit cases, said that it is completely a political issue. The forest office, along with *jote* land owners want to conduct the entire procedure under their authority but since the issue of land is related here, the involvement of DC's office is mandatory.

On the other hand, Hironmoy Chakma the president of *Upajatiyo Kath Babshayee o Jote Malik Kalyan Samity* pointed out the fallacy that the DC's office believes in and the confusion between the words 'verify' and 'inspection'. According to Sub-rule 4 (d) of Rule 4 of the Chittagong Hill Tracts Forest Transit Rules, 1973, the DC's office is entitled to verify the records of the applicants for *jote* permit and a range officer is authorized to inspect the site and verify the quantity of timber for which the extraction permit has been asked for. In practice the DC gives the authority to the TNO to verify the papers and inspect the *jote* land. However, it needs to be mentioned that the TNO has no records in his office to verify if an applicant is genuine or not. Hironmoy Chakma stated that the entire process of *jote* permit is time consuming and rigid due to bureaucratic complexities; the entire process generally takes almost three years. The DC's office allegedly harasses the *jote* owners by imposing many fabricated rules for inspection of the *jote* land whereas according to the transit rules they are authorized only to check the authenticity by verifying the existing documents in the DC's office of that particular *jote* owner. Alleged bribery is ubiquitous in this entire process. The amount ranges from Tk.3,000 to Tk.10,000 for the initial process of a *jote* application at the DC's office based on bargaining power. Again for each inspection and investigation the *jote* owner has to allegedly bear a cost of Tk.20,000 as bribe to ease and release their papers fast from the TNO.

The *jote* permit is supposed to be free under the Transit Rules. According to a source connected to timber trade a *jote* permit costs dearly. Each cft of timber costs an extra cash of Tk.100 to Tk.200 at the time of issue of a *jote* permit. Another Tk.200 is paid in bribe at different stages up to Dhaka, says the source. As a result the net profit of *jote* owners is on the decline and they are discouraged to invest their time and money in production of tree.

The deputy commissioner of Rangamati Hill District Mostafa Kamal blamed "unsuccessful afforestation programs and weak management system of reserve forests" as main factors for deforestation. He said the timber is extracted at a faster rate than the plantation rate. He also pointed out that Forest Department was asking for more land to be declared as reserved forest whereas most of the current reserve forest is empty of large to medium trees.

Md. Saidul Haque, a source at the Chittagong HT South Division office informed that on an average everyday 700 cft of timber is extracted from Rangamati South Zone. He informed that the extraction rate is higher during the monsoon season as transportation of timber by waterway is easy and cheap. Every month, on an average 25 to 30 applications for *jote* permit are submitted in his office; 15 of them are given. When asked about the selection process, he informed that there is no specific reason for that. The size of maximum *jote* land ranges from two to five acres and the biggest size goes up to 15 to 20 acres, which are actually rare.

The delay and hassle in issuing *jote* permits greatly frustrates the *jote* owners and they feel discouraged to grow trees. Failing to secure the expected cash from tree farming in time they explore other income sources. To explain the exhaustive process to secure a *jote* permit, Annanya Shadhon Chakma described the misery of a lady, an owner of a *jote* land. Due to an accident she became paralyzed and was unable to work. For treatment she needed cash, which she wanted to get from the sale of her timber she grew. She applied for a *jote* permit but the process was so long that she did not get the money when she truly needed it. Even being the owner of a property she failed to utilize it when she needed it most. This story clearly depicts the reason why *jote* owners are discouraged in investing in planting trees.

Fluctuation in estimating the standing trees causes another wave of deforestation. Suppose the estimation of a standing tree is 10.5 cft. After the harvest the timber is found to be 3 cft short of the estimate. This compels the *jote* owner to make up the deficit from somewhere else (i.e. reserved forest). This has a cost. For each extra cft of wood there is an alleged bribe. Sometimes, the *jote* owners are forced to take a permit for the extra cfts of wood whether they are willing to take it or not. Such permits for extra wood costs extra cash. This is bizarre, but the jote owners are compelled into it to secure a much needed jote permit.

Goda jam, garjan, champakul, badi, gutgutia, etc. are called *lali kath* because these timbers are reddish. The Forest Department has banned extraction of many species, which are almost extinct. No permit is given for these trees that have tremendous ecological value. But in reality these trees are harvested and illegally transported. Such damaging of the harvest needs to be controlled urgently.

According to transit rule there are supposed to be 24 check posts in the entire CHT region but in reality there are more. For example, from Kassalong river to Rangamati one has to go through Marishabit check post, Bablakhali check post, Mahilla check post, Mainimukh check post, and Kasalongmukh check post before finally reaching Shubalong Bazaar. In each check post the traders or the *jote* owners transporting the timbers have to pay a hefty toll for each cft of timber. Shaon Farid, General Secretary of Rangamati Kath Baboshayee Samabay Samiti (Rangamati Timber Traders Cooperative) Limited informed that each truck of timber cost them 20,000 Taka excluding the labor, transport, and timber's original cost and this cost is increasing day by day at an exorbitant rate. The *jote* owners and timber traders strongly believe that if the Forest Department performs its duties efficiently it would deem the other law enforcement agencies unnecessary. They argue that the forest department is aware of all sorts of timber extraction and transportation activity in Rangamati regardless of their legality.

Wastage at the start of the timber extraction is significant. The laborers engaged in extraction try to make their task easier by logging the trees some 3 feet above the ground. This results in a lot of timber being wasted that could have otherwise been put to use had the workers started logging from the very roots of the trees. The workers moreover swipe the trees 5-6 inches with the circumferences of 2.5-3 feet to test their maturity. A significant percentage of trees are found premature for extraction. This results in the premature demise of many trees.

To reduce transportation costs and to save space in the trucks the timber is cut into cubic shapes. The cubic shaped piece of timber is called *rodda*. This activity also results in the wastage of timber.

The wood smugglers backed by politically influential people always target the reserve forests in which they see mature trees that are superior in quality than the wood that comes from *jote* land. This has led to rapid destruction of the reserved forests.

According to sources in Uapjatyio Kath Baboshayee O Jote Malik Kallyan Samiti there are approximately 1300 *jote* owners and traders in the Rangamati Hill District alone.

Hub of Furniture Lovers

Rangamati town, a kind of capital of the CHT, is a hub of the wooden furniture lovers of the country. Huge quantities of teak and a few other timber trees are sawed everyday to feed the furniture factories. There are an estimated 1,500 furniture making factories and show rooms in Rangamati alone. A few thousand Bangali laborers and craftsmen stay busy day and night working almost non-stop. Thanks to the Hindi songs that keep them charged and sleepless at night.

Everyday these factories process some 7,000 cubic feet (cft) of wood, most of them teak, to make furniture and most of these furniture goes out of Rangamati. According to Hironmoy Chakma, most of these furniture shops and factories are illegal.

The furniture wood (mainly teak) used in these furniture shops is also harvested, sawed, and traded illegally. A portion of round wood procured through *jote* permits gets into the sawmills and ultimately to the furniture shops. This makes the furniture, made and bought in Rangamati, much cheaper than those made with the same wood in Dhaka or elsewhere. This is an incentive for illegal trade. The forest department, DC's office and others

A carpenter busy making furniture in a factory. There are hundreds of factories in Rangamati town making furniture day and night.   © Philip Gain

in the administration keep a blind eye about this illegal trade because they all allegedly gain significant financial benefits. The destination for timber furniture are mainly Dhaka and Chittagong.

The Chittagong Hill Tracts Forest Transit Rules, 1973 does not allow any furniture trading outside the territory. Then the question arises how such large quantities of furniture manage to go out of the district? In this case there exists a very preposterous game of plan and rule. It is the Furniture Transport Policy 2002 that allows the Government and semi-government officials to make and transport furniture out of CHT. The rule allows the transport of furniture out of the CHT for the purpose of wedding gifts. This transfer of furniture requires a transfer permit from the DFO of the concerned area. It should be mentioned that as evidence of a wedding gift a mere wedding card is attached with the application and that is considered as sufficient evidence to grant a permit.

The Furniture Transport Policy 2002 requires a trade or dealing license from the DC's office that needs to be renewed every year. Both the new license and renewal require a recommendation from the forest department.

Round wood including *lalli kath* (reddish wood) are coming to Rangamati town for sale.   © Philip Gain

This policy also specifies the quantity of wood that can be used for each particular furniture item and also mentions the permissible quantity of furniture for officials of different class.

When someone goes to the CHT for employment s/he submits a list of furniture to the DC's office. The DC's office sends the list to the forest department's office. When an employee is transferred out of the CHT s/he can take back the furniture (brought in and/or used) with a clearance from the DC's office. The forest department does the necessary inspection. There is a forest committee in the DC's office to oversee.

It is only the used furniture that can be taken out of the CHT. In the documents that allow the furniture to be transferred, these are used. But in reality one will hardly see any used furniture transferred. All the furniture seen transferred are new. "For illegal manufacture and transfer of furniture bribing is common. Different bribes are fixed for different items," said a local source on condition of anonymity. "The forest department official(s) assigned for inspection always gives a false report."

Saw Mills: There are 25 sawmills that have license from the Forest Department. The Deputy Commissioner's office also issues license for sawmills. It has a different list of sawmills with names of eight sawmills not included in the list of FD. The DC's office in its list has names of 12 sawmills common with the FD's list. These sawmills of Rangamati are spread over Rangamati town, Chandraghona, Kaokhali, Manikchhari, Rajasthali, Kaptai, Shubalong, Barkal, etc. areas. There are reportedly many illegal sawmills (50 according to a source) in Rangamati Hill District. The sawmills allegedly get supplies of teak and other timber trees in violation of the forest transit rules.

The manufacture and trade in furniture, illegal in most part, is possible for a syndicate of people in the offices of the DC, FD and also in the security agencies, which for financial gains, make the issue *jote* permit difficult and complex. The consumers of furniture do not care about what happens, where the timber comes from. All they care about is cheap teak furniture. This makes the *jote* owners to sell illegally their timber to the sawmills at a much cheaper rate. The end result is the continued destruction of the forests leading to erosion of the soil, water sources and many other ecological troubles for the CHT.

Tale of A Teak Gardener

Dibakar Dewan (60), a resident of Rangamati town, owns a *jote* of five acres in 107 Baradam Mouza. He bought the land in 1991 for Taka two and half lacs and spent Taka one lac to raise the garden. He invested his hard cash in the hope of good profit decades later. An employee of Bangladesh Agriculture Development Corporation (BADC) for 22 years he took the 'golden handshake' and received a cash of Taka five lacs and fifty thousand.

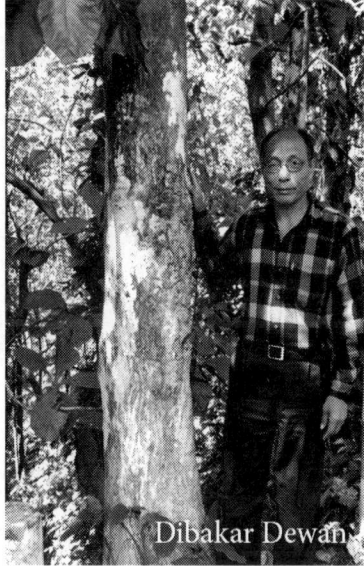

Dibakar Dewan

Dewan expects to harvest of the timber from his jote by 2025. He planted 350 teak saplings. The growth of the teak looks good. He has already sold four trees, each for 4,000 Takas.

Dewan knows that teak monoculture is not good for the environment and no other plant grow with teak. "I still preferred teak to fruits because fruit gardening requires intensive care, which is not possible without living close to the garden. Teak gardening does not require any close supervision. Once planted the saplings can be left alone to grow," says Dewan. "After the harvest of the teak the severe soil erosion becomes inevitable on the slopes and no other plants can be grown on the slopes that has once been planted with teak."

Dibakar Dewan is one of the thousands of indigenous persons in the Chittagong Hill Tracts (CHT) who have no other choice but to engage in growing teak and a few other timber trees on the land they still hold. Planting trees is necessary for their survival. The timber they grow is also much desired by the timber and furniture traders and ultimately the consumers far away. But the hassle the small jote owners face is endless!

by Philip Gain.

Water—Scarce in the CHT

Tahmid Huq Easher

Since the ancient times, the natural water sources like rivers, canals, natural springs, dug wells and hilly holes in the CHT area were believed to be "Goddess" and have great religious values among the ethnic communities. They established their habitat near the natural water sources in the hilly area, which is also good for *jum* cultivation. They used to drink directly from the natural sources as they consider the water from these sources as the blessing of the Goddess and believe that misfortune will come upon them if they pollute or disrupt the water sources (NGO Forum, 2008). Then gradually the population has increased over time in the CHT and man-induced disturbance to nature has increased in this area as well. Due to these cumulative impacts, natural water sources are becoming drier and contaminated, which lead water scarcity to be a prominent problem in the CHT.

Water Sources

Surface water: Water resources in the CHT comprise surface flows from the main rivers systems namely, Karnaphuli, Chengi, Maini, Sangu, Matamuhuri, Feni, Reingkyong and Bagkhali with a total length of 1,400 km (ADB, 2010). In addition to perennial rivers, upland communities rely predominantly on relatively shallow subsurface flows from the local springs. The monsoon rain is the main source of surface water in the region. After heavy rainfall excessive run-off occurs through the hills as a form of springs and falls into natural water streams like river and lake and hand dug wells. During the rainy season the river level remains high but during the dry season the level becomes low. Approximately one third of the population in CHT relies on water from wells and 17% drink water from springs, rivers or ponds (ADB 2010).

Traditionally female members of the family take the responsibility to fetch water from the streams whereas most of the surface water sources are remotely located. In the rural part of the CHT, women spend at least two hours a day to collect water from an average three kilometers distant water sources (Green Hill 2008). They also dig wells besides the streams. Water accumulates into the wells from the streams and the women fill up their pitchers (NGO Forum 2010). But it takes a lot of time to take water from such wells. They take baths and wash their cloths at the natural springs and streams. They also use the same water sources for drinking and cooking.

Ground water: The geology of the CHT is complex and is characterized as series of folded tertiary formations, which make the area unfavorable for extensive groundwater (NGO Forum 2010). The aquifers in this area have low transmissibility, which incur a large drawdown. Few wealthy families have developed wells successfully on individual basis for water source in their land. They usually use ground water for drinking and cooking.

Potable water sources: People in the urban part of the CHT use supplied water, which they get from local municipalities. The rural people living in the hills get potable water from shallow tube-wells, concrete water reservoirs, ring-wells, hand-pumps, deep set pumps, gravity flow piped water systems and infiltration galleries. Due to hard bedrock underlying much of the upland area tube well installation at greater depths is difficult. Only the rich families can afford the cost of a better potable water system from ground and surface water. The government and non-government organizations have also established several potable water systems for the rural community. Less than half of the population in CHT areas has access to potable water (ADB 2010).

Water Source Challenges in CHT

The physiography,   settlement pattern,   poor communication and remoteness of the CHT make the access to safe water source really difficult. During the rainy season, the arduous paths in the hilly area become extremely risky and slippery. It worsens the situation to fetch water from distant sources whereas it is already a physically demanding responsibility.

Over the past two decades there has been a noticeable depletion in spring flows resulting in significant impacts on rural communities. The CHT watersheds have been continuously degraded due to destructive activities such as unregulated timber and bamboo harvesting, improper road alignments and construction, and soil exhausting root crop cultivation in steep to very steep slopes. According to ADB (2010), widespread degradation of natural dense forests, big changes in land use, changes in climate, and unplanned physical interventions (such as road) are the major underlying causes for deteriorated spring flows in the CHT. These result in the drying of water sources in the rivers, and lakes, and accumulating siltation at river beds and lakes, which deteriorate the water quality in this area. Also an overall decreasing rainfall pattern has been observed in six of the seven stations within the CHT during the last decade, which worsens the situation of surface water sources.

In the rainy season landslides make the surface water turbid and create on abundance of organic and inorganic matters. Also in this season, the water streams receive agro-chemical, domestic wastes, and silt though surface runoff. These factors cumulatively restrict the surface water use in the CHT during the monsoon.

The rural people in the CHT are unaware about the surface water contamination with microorganism, pathogens, and pollution. They use contaminated surface water for drinking and cooking without boiling or treating. Thus they suffer from water related diseases like diarrhea, dysentery, and cholera.

In the dry season when the natural streams become drier, local people depend more on ground water. But the hydro-geological conditions, high altitude and, high implementation cost and complications to install tube wells limit the use of ground water (NGO Forum 008). The geo-physical condition of the CHT doesn't allow installing easy and low-cost technologies. Advanced technologies like ring wells, deep set pumps, gravity flow piped water systems and infiltration galleries are being introduced in the CHT; these options, however, are highly expensive and unaffordable for the poor indigenous communities. These options are also neither suitable in all parts of the CHT nor operational round the year (NGO Forum 2008). The operation and maintenance are also complicated and it needs skilled person to fix it. Without proper training these technologies won't be fully functional and will be left inoperative (NGO Forum 2010). Only rich communities can afford these techniques. Governmental and few non-governmental development agencies have installed some ring-wells and gravity flow piped water systems in few a communities. But the poor people have less access to these sources as the rights to use water from these sources are blocked and manipulated by the rich communities (NGO Forum 2007).

Sanitation in the CHT

The indigenous people of the CHT don't have any proper sanitation system. They usually defecate in open spaces beside the hills or in the bushes. But they never defecate in or beside the natural water bodies as they worship these streams as Goddesses. They traditionally use leaves for cleaning after defecation. Only a small percentage of the CHT people are used to the system of using a latrine (pit latrine, plastic pan etc.).

The overall sanitation coverage is very poor in the CHT and over 80% of the households use open latrines or open spaces (ADB 2010). The sanitation coverage is Khagrachhari, Rangamati and Bandarban is 57%, 54% and 14% respectively, which is lower than the national coverage (Ali 2010). Only 7% people in Rangamati, 2% in Khagrachhari and a mere 1% in Bandarban practice safe hygienic behavior, e.g. washing after defecation (Ali 2010).

The poor ethnic communities who live in the remote area have minimum ideas about safe water, sanitation and hygiene. In most cases, these communities are not oriented about the necessity of using safe water and a proper sanitary system and, maintaining hygiene behaviors. They are completely unaware about the relation between water borne diseases and poor hygiene practices. Most of the indigenous communities have inadequate knowledge on children defecation practices (defecating or disposing in the latrine). They don't consider children feces harmful for their health (NGO Forum, 2008). Most of the households don't cover or bury children feces properly in specific holes or household latrines. Most of them also don't wash their hands properly after cleaning children feces.

Many communities use an indigenous technique called *Machang* as sanitation system (NGO Forum 2008). They make a platform a bit higher from the ground. They sit on the *Machang* for defecation and allow the stool drop on the ground. The pigs immediately devour this. According to this system is a natural recycling system but in reality it is an unhygienic system. The feces pollute the environment, increase mosquitoes and spread germs through the pigs and other domestic animals and birds.

The lack of appropriate knowledge on sanitation is the main reason for poor sanitation and hygiene situation due to poverty, natural hazards, water scarcity, poor communication, culture, practices and social beliefs of the CHT indigenous communities. The weight of materials for sanitary latrines and poor transportation system in the hills make it difficult to establish such latrines in remote areas of the hills. Also poor ethnic families cannot afford the cost to establish a proper sanitation system (NGO Forum 2010). Many of them use pit latrine, which is cheap to install but unhygienic. Deforestation, drying up of natural water sources, unplanned dam and cultivation, lack of administrative support and communication problem are also the major obstacles for the development of safe water and sanitation facilities in the hilly areas (*The Daily Sun* 2009).

The availability of health services is inadequate and poor. The government hospitals and clinics are close to sub-district headquarters whereas for 65% of the households in CHT areas the distance between these facilities are on average 5 kilometers (ADB 2010). But most of the time the doctors are absent in these health centers and only a handful of nurses run the whole center with their limited knowledge about medicines. The indigenous people mostly depend on the traditional doctors and traditional folk healers for the treatment of the diseases. There are a few health centers established by NGOs, which somehow provide treatment and consultation to improve the health condition of the local inhabitants.

The government and non-government organizations have tried to provide safe water and proper sanitation systems for the rural CHT people. But they didn't pay much attention to addressing WatSan issues right after signing the Peace Accord. International development organizations like UNICEF and, several NGOs like NGO Forum, Green Hill have been implementing several water and sanitation programs to support the communities to strengthen and develop safe water supply and sanitation facilities. In a few communities, NGOs helped establish proper sanitation system (plastic pan, concrete latrine, etc.) whereas the local families paid only 10% of the total cost. They are also working to improve the behavior in hygiene practices in the remote rural areas of the CHT (Ali 2010).

Actors and Actions

The government and different non-governmental agencies are working to

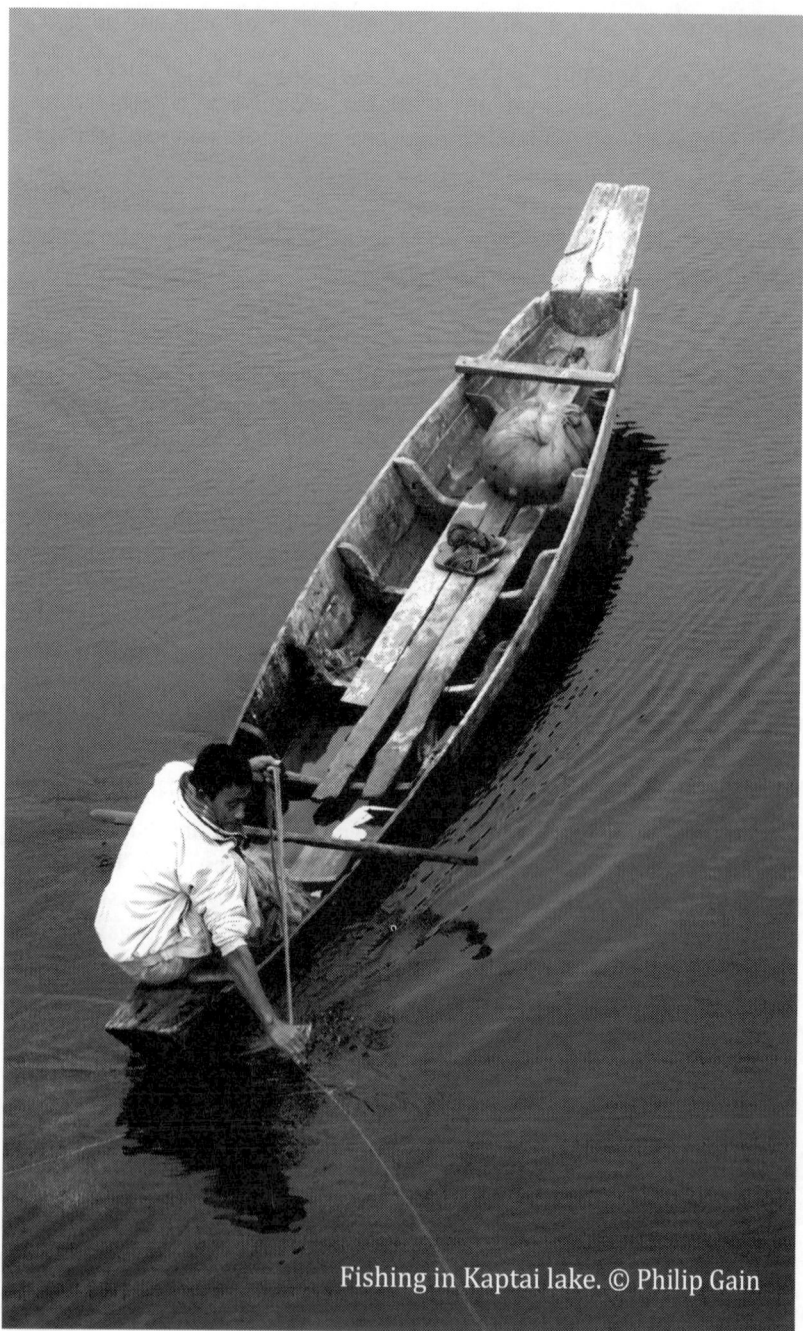

Fishing in Kaptai lake. © Philip Gain

improve the situation of safe water and sanitation in CHT areas. After the signing of the Peace Treaty, NGOs have started their programs to improve the situation of CHT communities. But initially they did not pay much attention to WATSAN (Ali 2010). International development organization like UNDP and UNICEF and, NGOs like WaterAid, NGO Forum, GoB-Danida, Dhaka Ahsania Mission, and Green Hill are working together and separately on the socio-economic and cultural issues and on developing the safe water sources and the sanitation situation in the CHT. In recent years, both government and non-government organizations are focusing more on WATSAN issues to improve the situation of CHT communities.

Government of Bangladesh (GoB): The government tried to provide safe water options in CHT areas by installing hand-pumped shallow tuebwells, but adverse rocky and stony ground hindered the process (NGO Forum, 2010). Later GoB has tried to install ring-well in few communities but did not train the local NGO partners and community to install, use, and maintain it properly. Thus lack of orientation and improper installation made the technology a failure.

However, since the formation of the Ministry of the Chittagong Hill Tracts Affairs (MoCHTA), a project titled, "The Integrated Community Development Project for Chittagong Hill Tracts Area" has been taken up for the overall development of the CHT (MoCHTA, 2011). The project has been operating since 1996 and in 2010 the second phase of the project has been completed. According to the website of MoCHTA, the government has developed 375 natural water sources and distributed 11,100 slab latrines in different communities in the CHT through the 2nd phase of this project. It has also trained 3,500 local workers on safe water use, proper sanitation installation and maintenance, and healthy hygiene behavior. These local workers are spreading their knowledge among their neighbors to improve the WATSAN issues in CHT areas.

UNICEF: It works with the Government of Bangladesh on intensive hygiene education, sanitation and, water supply and quality improvement projects in CHT areas. These projects are devoted to raise awareness about the importance of proper sanitation and to improve the sanitation and hygiene behaviors of the local community in the CHT. These activities target rural and urban communities in the three districts of the CHT (UNICEF 2008).

UNICEF works to improve the hygiene behaviors, sanitation practices and water supply points of local communities through awareness programs. It recruits the local workers as community hygiene promoters (CHP) in the CHT. Generally local young women are trained as CHP who educate their neighbors about the health benefits of sound hygiene practices. They facilitate discussions about sanitation and organize events to promote good hygiene within their communities through home visits, school visits, courtyard meetings, group meetings, tea stall and grocery shop sessions, video and film showings, and dramas. Also CHPs discourage people from using open and other unsafe water sources in their awareness sessions.

UNICEF and its partners have also been assisting in installing latrines and safe water sources for the local schools and remote communities in the CHT. It has helped many communities to improve their safe water supply, with a specific focus on arsenic mitigation. UNICEF addressed the naturally occurring arsenic contamination of groundwater by testing more than 1 million tube wells, providing alternative safe water in 68 upazillas and implementing public information and awareness campaigns on arsenic mitigation (UNICEF 2008).

NGO Forum: The NGO Forum for Drinking Water Supply and Sanitation has started its WATSAN activities since 2004 in the CHT. It has undertaken several community-managed water and sanitation programs in the CHT. It has been implementing programs to support the development of sustainable community-managed and community-owned water and sanitation programs. These programs strengthen the capacity of the local community in order to provide and maintain safe water supply and sanitation facilities, and encourage behavioral change in hygiene practices in the underserved rural areas in CHT (NGO Forum 2010).

With the local NGO partners, the NGO Forum has promoted different alternative technologies at community level to ensure safe water sources for the CHT communities. Till now with the local NGO partners, NGO Forum has installed 63 ring-wells in different communities in the CHT. Each ring-well can serve 20 households for their daily drinking and domestic uses. In 2010, NGO Forum has installed 150 concrete reservoirs in different communities in the CHT. Each reservoir can contain 3,200 liters rainwater, which serves the community people's consumption round the year (NGO Forum 2007). It has also assisted many local ethnic communities to develop their existing rainwater harvesting systems (RWHS) as a safe water source. It has introduced both household-based and community based RWHS systems, which contain 3,200-3,500 liters and 35,000-50,000 liters respectively at a time (NGO Forum 2007). Community-based RWHS can serve 20 families for six months as a source for drinking water. Already

NGO Forum has installed 65 RWHS in different communities of CHT.

NGO Forum has also offered operation and maintenance of the alternative technologies. It has selected caretakers (both male and female) from the local communities and trained them about the operation and maintenance of the technologies to operate the technologies properly for a long time. It has also formed Village Development Groups (VDG) to deal with the address safe water crisis, to introduce community management, to empower the community people in decision-making and to make the WATSAN program sustainable in the CHT (NGO Forum 2007). The VDGs have encouraged participation of the poor and the women along with the civil society people. The VDGs have created awareness among the communities in safe water use, proper sanitation practices, and healthy hygiene behavior.

It has also assisted few communities to install plastic slabs and rings for proper sanitation. These plastic slabs are cost-effective and can be easily installed in the hilly areas. It takes only 10% of the total cost to install this sanitation system from beneficiary families (NGO Forum 2010).

Green Hill: Green Hill (GH) has been actively implementing its "Sustainable Rural Safe Water Supply and Sanitation Programme in CHT" since 1999 with the support of WaterAid Bangladesh. It is a CHT based local non-governmental development organization. GH has enhanced community empowerment through the formation of Village Development Committee (VDC). GH has provided capacity building training on WATSAN issues to VDC members. Over the last few years, VDC members have utilized their knowledge in their own community and have succeeded in sustaining the WATSAN program in their area. VDC members are also getting involved in conducting surveys and assessment, monitoring necessary operations to address the WATSAN situations and to educate the local people about proper sanitation and safe water sources along with the GH representatives.

GH has constructed concrete reservoirs to preserve spring water in a few CHT communities. Each reservoir is channeled with plastic pipes from the natural springs and can hold 18,000 liters of water. It has also piloted the concept of 'ecological sanitation' in the Chittagong Hill Tracts. GH has installed several ecological sanitation latrines in different communities. The human waste from the latrines, mixed with soil and ash after each use, and urine, collected in separate plastic containers, can be used as a natural fertilizer (WaterAid 2010). The local people were trained to use and maintain the latrines properly. Now they use the composed feces as fertilizer in potato and bean farming.

DANIDA: DANIDA has been working with Government of Bangladesh (GoB) in the CHT since 2009 in safe water supply and sanitation. It has drilled 591 (both deep and shallow) tube-wells and 172 ring-wells in different communities of the CHT. It is also monitoring these artificial wells time to time and renovating and re-activating when needed. GoB-DANIDA representatives are also motivating the local community people for proper sanitation system, safe water sources, process of installing ring-wells, hygiene promotion, etc. through 1,130 community-based volunteers and 111 union facilitators.

Conclusion

Despite efforts by the communities themselves, the government and non-government organizations, the common people in the CHT continue to face the challenges of safe potable water supply and a proper sanitation system. Poverty is a big barrier for the hill dwellers in the CHT to avail their needs of safe water and proper sanitation. Developing ownership of the infrastructures of the on-going water and sanitation facilities installed by the government and non-government organizations in the region is the great challenge for the beneficiary communities. Affordability to operate and maintain the technologies introduce is another challenge due to the fact that the national and international funds and donations are limited and not properly distributed. Lack of coordination among actors remains to be a standing problem.

However, the government and non-government agencies are slowly and steadily introducing small scale but effective WATSAN projects and technologies in the CHT. Many people are benefitting from these projects and are scaling up their capacity of proper water and sanitation practices.

Bamboo Plantation for Riverbank Protection and Stabilization

Every year, riverbank erosion of River Lu Leng Chhara in Bandarban, threatens the lives and property of the people of four villages clustered closely together. In the rainy season the river water, due to landslides, becomes turbid and abundance of organic and inorganic matters in this area. Also in this season, the river receives agro-chemical, domestic wastes, and silt though surface runoff. As a result, the villagers have noticed a quality deterioration of the river water.

Bamboozled the villagers decided to plant bamboo along the riverbank to improve the water quality and reduce the impact of riverbank erosion. They selected two acres of land along the riverbank and planted *Borak Bansh* that grows well in the CHT and is relatively easy to cultivate. Its cultivation is also highly profitable. They maintained the contour system of planting and planted the bamboos at the lower slopes of the hill. They prohibited pulling of bamboo shoots. Community members take turns in monitoring the growth and maintenance of the young bamboo. They only harvested and sold matured bamboo. In the next six years, riverbank erosion in this area reduced significantly. Thus they increased the area for bamboo plantation. They also earned Tk.50,000 to Tk.70,000 annually from the sale of bamboo and used the money to repair the monastery or build schools for the community. This simple action created a positive impact on both the environment and socio-economic condition in that community. (Source: UNDP-CHTDF 2005).

Community Managed Water Supply System

The hill people of Kukimara village (Wagga Mouza, Kaptai, Rangamati Hill District) used to drink water direct from a nearby canal. But many people, especially children, suffered from diarrhea due to poor water quality of the canal. Then they were alerted by a community worker about the unhygienic state of the canal, which may have been the reason.

Shwe Prue Marma, the *karbari* of the village, discussed the matter with his community and located a permanent stream atop a hill near their village, which could be an alternative source of clean drinking water. But the community lacked the infrastructure and support. Then the *karbari* contacted Green Hill, a local NGO that provides a gravity flow water supply system. Unlike tube-wells and ring-wells, surface water systems (e.g. gravity flow water supply system) is more acceptable to the community and the technology is more appropriate and culturally adaptable. It is easily replicable and cost effective.

Green Hill agreed to support and organized a meeting with the community about the management of the project. Two reservoirs were constructed under the supervision of Green Hill. A large one was built at the principal water source, connected by guided gravitational flow to

Drinking water from a bamboo glass. © Philip Gain

another reservoir located just above the village. Seven supply points were also established. The community contributed 5% of the total cost. The community members collected materials (stone and sand) from their area and volunteered labor. Other materials were purchased from Chittagong with the help of Green Hill.

The construction of two reservoirs and seven supply points cost Tk.400,000 (approximately USD5,000). A village development committee (VDC) was formed to maintain and manage the water supply. Green Hill also trained a caretaker to handle routine repairs and for any major fix, an expert from Green Hill pays a visit. The VDC advises the community not to cut trees near the stream in order to preserve the watershed and to reduce soil erosion. The community faced some minor difficulties (cash contribution and collection of firewood) but since construction they have a steady supply of potable water. Women spend much less time walking to any water source. The incidence of waterborne diseases has also decreased. They also understand the value of preserving the forests around the stream, which also improves the socio-economic condition of the community. The Chittagong Hill Tracts Development Board (CHTDB) and Green Hill continue to offer support for the project. The Green Hill's representative still periodically visits to educate the community members on water and sanitation issues. (Source: UNDP-CHTDF 2005)

A Chak woman collecting potable water from a ditch.   © Philip Gain

Work Cited

ADB. 2010. *Chittagong Hill Tracts Study on Potential for Integrated Water Resources Management: Promoting Effective Water Management Policies and Practices – Phase 5.* Dhaka.

Ali, A. 2010. *Reaching the Hard to Reach.* Dhaka: NGO Forum for Drinking Water Supply & Sanitation.

Green Hill. 2008. *Changing Lives Through Appropriate Water Supply in the CHT.* Dhaka: Green Hill.

HDRC. 2009. *Socio-Economic Baseline Survey of Chittagong Hill Tracts for Chittagong Hill Tracts Development Facility.* Dhaka, Bangladesh: UNDP.

MoCHTA. May 5, 2011). *The Integrated Community Development Project for Chittagong Hill Tracts Area-2nd phase (2nd Revised).* Retrieved 4th April 2012, from Ministry of Chittagong Hill Tracts Affairs: http://www.mochta.gov.bd/index.php/activities/projects/210-qthe-integrated-community-development-projetc-for-chittagong-hill-tracts-areaq-2nd-phase-2nd-revised.

NGO Forum. 2007. *Serving the Unserved: Consolidation of Watsan Success in CHT (Volume 2).* Dhaka, Bangladesh: NGO Forum for Drinking Water Supply & Sanitation.

NGO Forum. 2007. *Serving the Unserved: Consolidation of Watsan Success in CHT (Volume 1).* Dhaka, Bangladesh: NGO Forum for Drinking Water Supply & Sanitation.

NGO Forum. 2007. *Serving the Unserved: Consolidation of Watsan Success in CHT (Volume 3).* Dhaka, Bangladesh: NGO Forum for Drinking Water Supply & Sanitation.

NGO Forum. 2008. *Study on Sanitation and Hygiene Situation in Chittagong Hil Tracts (CHT), Bangladesh.* Dhaka, Bangladesh: NGO Forum for Drinking Water Supply & Sanitation.

NGO Forum. 2010. *Healthy Living in the Hills.* (J. Halder, Ed.) Dhaka, Bangladesh: NGO Forum for Drinking Water Supply & Sanitation.

Progotir Pathey. June 2009. *Monitoring the situation of Children and Women: Multiple Indicator Cluster Surver 2009.* UNICEF, BBS, Progotir Pathey. Dhaka: UNICEF.

Hard to Reach Areas: Natural Disasters Hamper Access to Safe Water - Speakers Tell Consultation Meeting. *The Daily Sun,* July 1, 2009.

UNDP-CHTDF. 2005. *Chittagong Hill Tracts Best Practices Handbook: Promotion of Development and Confidence Building in the Chittagong Hill Tracts.* Dhaka: UNDP.

UNICEF. July 2008. *Rural Sanitation, Hygiene and Water Supply in CHT Area, Bangladesh.* Retrieved April 2, 2012, from UNICEF: http://www.unicef. org/bangladesh/RURAL_Water_Sanitation_and_Hygiene.pdf.

WaterAid. 2010. *End of year report 09/10 – Enhancing Environmental Health through Community Organisations, Bangladesh.* Dhaka: WaterAid Bangladesh.

Stone Mining in Bandarban: Hill Villagers Go in Water Crisis

Shekhar Kanti Ray & Md. Safiullah Safi

G heraopara is a Marma village in Roangchhari Upazila of Bandarban Hill District. Eight kilometers south of the upazila town the village has 43 Marma families. Like elsewhere in the remote areas of the Chittagong Hill Tracts the hill streams are the main sources of water of the *jum*-dependent families of Gheraopara in the desolate geographical setting. The Gheraojhiri stream flows north to south passing closely by the villages. Another stream, Jhadijhiri, converges with Gheraojhiri from the east. Gheraojhiri has been the main source of water for drinking, washing, and bathing. The villagers even caught fish when stream had enough water.

Four years ago (2007) things started to change for the worse when some stone traders started mining stones from the Gheraojhiri stream bed and banks adjacent to Gheraopara. The stone miners had permission from the Deputy Commissioner of Bandarban. Ever since the mining of the stones, the water of Gheraojhiri started to decease. Worse, the water still flowing, got polluted and became undrinkable. The villagers resorted to harvesting potable water from Jadijhiri. But the stone miners are now (2011) trying to mine stones from this stream as well. Witnessing their sources of drinking water being increasingly lost, the villagers find themselves in deep trouble. If their drinking water sources are completely exhausted, they can no more live in the area.

Aung Thowai Ching (54) *Karbari* of Gheraojhiri village has his *machang* house of bamboo, wood, and straw just on the top of the confluence of the Gheraojhiri and Jadijhiri streams. "The water of the Gheraojhiri has noticeably dried up due to the stone mining. Now the stream is filled with filth and dirt. As a result we get our drinking water from the neighboring stream Jhadijhiri. But now the stone traders are conspiring to mine the stones from there as well. We are doing our best to prevent them. If the remaining stream is similarly destroyed, our subsistence will become very difficult".

Pointing to Jhadijhiri said, "Though this stream is relatively smaller than Gheraojhiri, the ware flows faster and the water is cleaner. It is so because of big chunks of stone half a kilometer from here. The stones keep the water remains pure due to the presence of the stones. The water flows faster due to blockage on the path. As we went atop Jadijhiri, we came across pebbles, uneven stream bottom, clearer water and shoals of small fishes–resembling the description given by the *karbari*.

The bottom of Gheraojhiri stream is plain due to the removal of stones. "Water here used to reach up to the level of the knees in the past. Now it has declined to half its height," informed the *karbari*. About a hundred yards north of the confluence of the streams, we found a group of women and adolescents of the village. Some of them were waiting for their dug up holes on the stream to be filled with water. They will proceed to take bath when sufficient fresh water is accumulated in the holes. Two women of the *para* (village), Mee Frue Ching (28) and Craw Awe (30) were busy separating the fish and the snails caught in the fish trap (made of bamboo slips). They informed us that the fishes were caught from the Jadijhiri stream. Upon further discussion they revealed that the fishes have drastically decreased in the Gheraojhiri; so now they catch fish from Jadijhiri. They also resorted to Jadijhiri when it came to the drinking water. They did not go beyond bathing and other necessary washing in Gheraojhiri due to its decadent condition.

Describing the condition of the stream prior to stone mining, Gheraopara's grocery shop owner Mong Ching Sui (50) said, "Even five years ago, it was possible to bring up about 100 bamboos all at once over the stream. Now the water is so much less that it does not permit even 50 bamboos to be picked together. While previously it took five to ten minutes to complete bathing, now it takes about half an hour. The habitat of the fishes has been destroyed due to the absence of the stones; as a result fishes are now hardly found in the stream." He expressed his unison with the stand of the *karbari* to prevent any attempts to destroy their only remaining source of drinking water, Jadijhiri.

One kilometer further in the south from Gheraopara, the stream Kanaesejhiri convergences with Gheraojhiri. Here the stone traders have amassed stones plucked from half a kilometer of Kanajjyajhiri. The tracks of the truck wheels were embedded in places. The *karbari* and the local people informed that the headman of Gheraojhiri *mouza*, Shai Sha Aung and a businessman from the Roangchhari Sadar, Rabi Sen Tanchangya were mining stones at Gheraojhiri, four kilometers north of the confluence of Gheraojhiri and Kanajjyajhiri. Rabi Sen Tanchangya said that they were simply utilizing the permit issued by the Bandarban District Administration to Kanchanjoy Tanchangya, a member of Bandarban District Council. The permit allowed them to mine 30,000 cubic feet of stones. When consulted Kanchanjoy Tanchangya confirmed that he had a permit from the Deputy Commissioner's office of Bandarban. However, Kanchanjoy Tanchangya himself was not directly involved with the mining. Rabi Sen Tanchangya and the headman of Gheraojhiri were mining stones by utilizing his permit. He had no idea how and many stones were being mined. But he acknowledged the fact that the unplanned mining of the stones is causing the decline of water in various parts of the stream.

Gheraopara Marma village is one of the villages hard-hit by water scarcity due to stone mining and the resulting destruction of water sources. The village still explores a solution for availability of alternative water sources and good communications with the upazila center. However, a remote Adivasi village with 10 Murang families (Amtoli Murang Para in Fashiakhali Union) without alternative water sources is faced with genuine trouble.

It takes almost four hours by jeep to Amtali Murong Para from Dulhazra Bazar of Chokoria upazila on the Cox's Bazar-Chittagong highway. Travelling through the hill forests, rubber, and acacia plantations (also known as hybrid plantation) one will not miss stockpiles of stones in the plains land of the Shaperghera area. Hla Mong Cha (26), a local Marma, informed that Khairul Bashar, a stone trader, has leased an acre of his land for five months to stockpile stones. Cha has no idea where the stones have come from and where these will go.

Passing Fuittajhiri, Shapmarajhiri, and Goyalmara, we finally reach Amtoli Murung Para. All ten families of the Murong village are *jumias* who had had good days so far. But the stone and wood traders have brought many troubles for them. "Nur Hossain, Ali Hossian, Nazrul Master, Afsar and others have destroyed our groves to build a road. They have not compensated for the damage, they have done to us," murmured Lai paw (30) a Murung of Amtoli. Karew Murung (40), a *karbari* and resident of Amtoli alleged that the stone traders have dammed the stream and spoiled its water by throwing branches of trees and mud in it. "This stream was our only source of water. In the past, it was a safe source of our potable water. But now the people of this village suffer from stomach ailments. Two of our boys have died from diarrhea. We believe they got germs from the water of the stream. We pass our days in fear. A family has already left the *para*. If the conditions do not improve, more are likely to follow," said Karew Murung. Kaiwan Murung (42) of the village reported she boils water before drinking.

Shyamchhara is one kilometer east from Amtali Murung Para. The canal is dammed in many places to allow trucks to move. Sand and mud fallen from the roads and branches of trees have reduced the water flow of the stream. Piles of stones on the riverbed are also noticed. The stone miners have built temporary houses on the roadsides.

A group of 15 to 20 workers were taking preparations for extraction of stones. Obaidul Haque (35), Jinat (15) and Abbas (20) were among them. They and most of their fellow workers came from Fasiakhali in Lama Upazila. They collect stones of all sizes from the canal. The supervisor of the group (*majhi* or middleman) Babu Miah (50) from Bogachhrai of Fasiakhali informed that they were collecting stones with a permit issued to Noor Hossain of Fasiakhali. This group has so far collected around 30,000 cubic feet of stones and have plans to collect another hundred thousand cubic feet.

Noor Hossian denied being involved in stone mining business in Shamchara. He claimed that he collected stones from Eidgahchhara a month ago. But he admitted that Babu Mia was his employee. The acting chairman of Fasiakhali union, Mohammad Kamal Hossain, informed that Noor Hossain is among three (the other two are Khairul Bashar and Anis) are indeed engaged in mining stones from Eidgahchhara and Shyamchhara. He acknowledged that stone mining from streams does obviously pollute water. However, stone miners dig some wells for potable water for the communities.

The experts fear environmental disaster and destruction of water sources due to the extraction of stones from the streams in Bandarban district. In a letter (dated 1 March, 2011) to the Director General of the Department of Environment and the Deputy Commissioner of Bandarban, Prof. Dr. Harun-ur-Rashid, director of the department of Environmental Science of Independent University Bangladesh (IUB) communicated his concern about the stone mining in Bandarban district. "The extraction of stone threatens to permanently put at risk the water supply of the whole district. The stone layers help in formation of perched aquifers, which keep the thousands of *jiris* (streamlets) flowing all the year long. By pulling out the lawyers of stones along the streams many of them have been dried out. This has forced nearby villages to be moved to new areas, creating severe competition for water sources. We should not allow the future of the district to be jeopardized for the sake of saving on construction costs through the use of local stones," wrote Prof. Harour-Er-Rashid.

Commenting that the streams get affected due to stone mining, Prof. Dr. Syed Shamsuddin Ahmed of the Department of Geology and Mining of Rajshahi University said, "The stone keeps the stream water clean. It prevents sedimentation in the streams, and consequently retains the depth of the streams. The mining of stones causes the filling of the streams from the mud falling from the either side of the hills. The faster the stone are mined, the faster the streams will be filled up." He suggested for quick actions against such destructive stone mining.

Deputy Commissioner of Bandarban, Md. Mizanur Rahman said, "Permissions have been given to mine stones from various streams in Bandarban. But mining can be done only by hand, and not by any machines that cause harm to the streams." He assured that he would look into the matter and take action against stone mining done by any means not permitted.

Stones harvested from streams are stockpiled in Shaperghera area in Lama.   © Uzzal Tanchangya

Houses in the CHT

Han Han and A.K.M. Muajjam Hossain

The mountainous landscapes of the Chittagong Hill Tracts (CHT) offers a rich and unique vernacular architecture that is remarkably different from other plain parts of the country. Unlike Bengali rural settlements of the flat land where houses are built around courtyard, all ethnic communities of the Hill Tracts build their houses on stilts known as *Machang* houses, strongly characterized by its post-lintel structure with bamboo wall panels and a thatched roof.

Since antiquity, the inhabitants of this hilly land have been building their traditional houses from a spontaneous response to their lifestyle, culture, environment and climate, using only local natural materials, simple tools and age-old techniques, depicting a unique process of formation of ecologically sustainable architecture.

Settlements

The hill people live in small villages or hamlets known as *paras*. Such settlements of the different ethnic groups may initially look as if they are scattered at random over the valleys, high mountains and deep forests. But a closer observation will reveal that each group usually has its own preference for a particular part of the hill region for a settlement. In their writings, Claus-Dieter Brauns and Lorenz G. Loffler recognized different ethnic groups as hill dwellers, valley dwellers and dwellers living in between.

"... the Khumi, the Tongcengya, the Hill Tippera, the Khyang, and, finally, also the Mru. These small ethnic groups are generally found in the transitional areas—that is, in small river valleys and on the lower crests of hills—between those unmistakable hill dwellers of eastern origin and those who are clearly valley dwellers. The hill dwellers include the Lusai, the Pangkhua and the Bawm. They are never attracted to the valleys and their villages are therefore nearly always found on hill tops and spurs of hills. The Chakma, the Marma and the Tippera are, on the other hand, valley dwellers who will settle in higher regions only when pressed for lack of land."(Brauns and Loffler 1990: 34).

Since the availability of a water supply is the primary consideration of any hill settlement, villages are usually found near *jhiris* (small water springs). Other than rivers, *jhiris* are the main source of water for drinking and irrigation in the hilly terrains. Most of the time *paras,* which have no proper name are called by the name of its spring.

Generally ethnic people do not shift their villages unless relocation of a settlement becomes necessary, such as a recurrence of mass illness or a scarcity of water, like a spring drying out. Throughout the past several decades though, land encroachment, urbanization, influx of Bengali community and loss of livelihood have compelled the hill people to leave their natural habitats. Some have come down to the flatlands or have taken refuge in remote mountain areas.

A *para* mainly consists of individual houses of each nuclear family. Only larger villages can afford to have additional structures like a Buddhist temple or church and one or two tea stalls. In recent times, villages with good access or nearby roads often include a primary school. Villages have one main entry and some secondary entries. As the villages are not enclosed or demarcated by boundaries, entries are identified by the pathways.

Size and pattern of a settlement varies for different ethnic groups. For example, there are generally 60 to 100 households in a Marma *para,* while in a Mru *para* there are only 5 to 20 houses. In a typical Marma *para, machang* houses are arranged in rows with their open terraces facing the circulation paths that run across the village middle. In a Chakma *para,* houses generally face east and in a Mru hamlet, there is no visible order of placing the houses. The headman who is known as *karbari* often settles at the centre of the *para.*

Houses

One obvious reason for building houses on stilts by all hill communities is to achieve a flat platform without leveling the hilly ground itself. Apart from that, stilt construction also safeguards the house from gushing water during monsoon season, allows under floor ventilation and secures the house from wild animals accessing it. Like other rural houses of the country, *machang* houses are pitch roofed and have no rain gutter, allowing the freefall of rainwater during monsoon season.

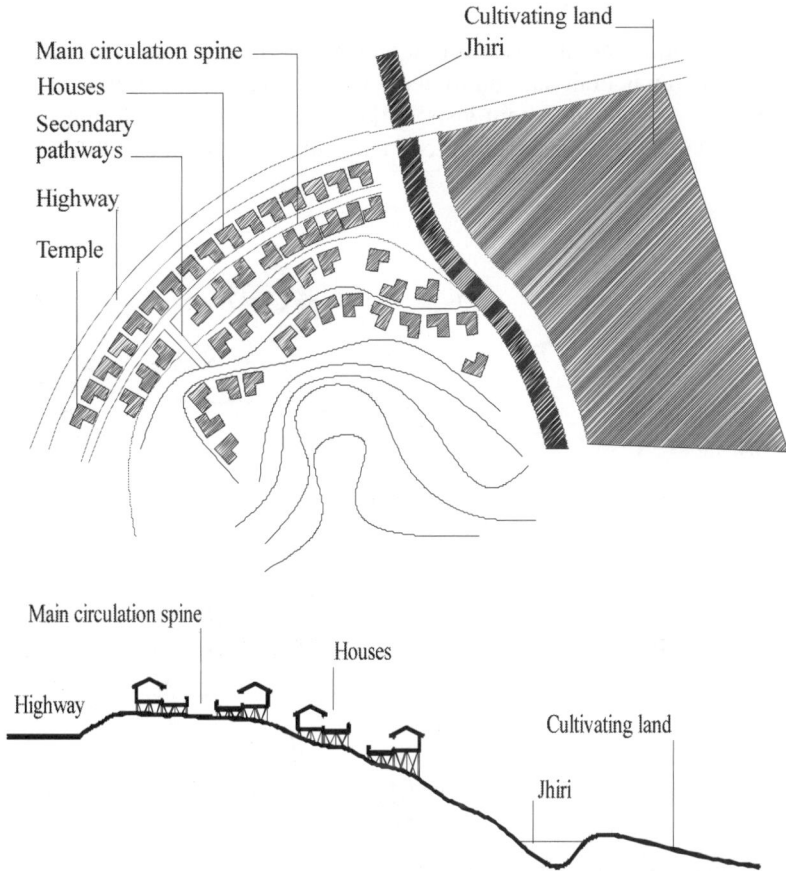

Main circulation spine

Houses

Secondary pathways

Highway

Temple

Cultivating land

Jhiri

Main circulation spine

Houses

Highway

Cultivating land

Jhiri

Plan layout and section of Hangthama rwa, Roangchari, shows the typical settlement pattern of a Marma village.

Although the indigenous groups living in Chittagong Hill Tracts have similarities in habitation due to geographic condition, they are culturally distinct from each other. The varied rituals, social rites and nature of social relationships have directly influenced the design of their dwellings. This cultural difference is so evident in the architecture expression of the dwellings that one can distinguish a particular indigenous group's house from that of the other indigenous groups by its form, discrete design and internal spatial organization.

For instance, a Tanchangya house is easily distinguishable by its diagonal wall.

"The unique characteristic is that one wall of the house in the main living area is built slanted to allow reclined sitting for relaxation. Indeed, in the settlement visited, there was only one such house left, which was thus documented." (Ahmed and Kabir 2005: 38).

Tanchangya house, typical plan and section.

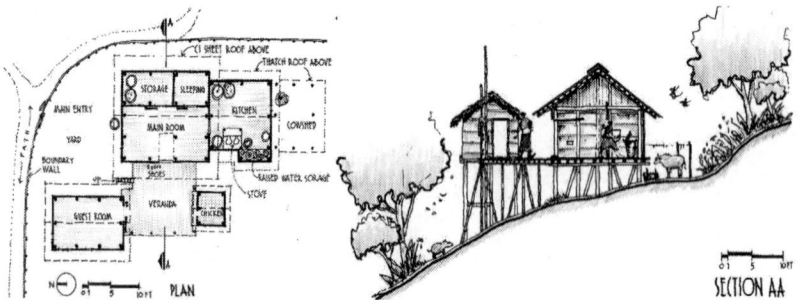

Khyang house, typical plan and section.

Source. Ahmed and Kabir 2005

Spatial Division and Use

Basically a typical *machang* house comprises of several rectangular blocks allocated for specific household functions. If required, the extension of an existing house is done by simply adding on another block to the structure. For example, a typical Marma house consists of three structurally independent square parts: a terrace, a main house and a kitchen.

1 alter
2 over hanged shelf
3 bed
4 shelf
5 elevated water storage
6 hearth
7 dining table
8 grain storage
9 pottery shelf

A Tamyang
B Praw
C Toi-the
D Kya

praw
toi_the entry
raised platform for water storage
toi_the
tamyang
plinth structure

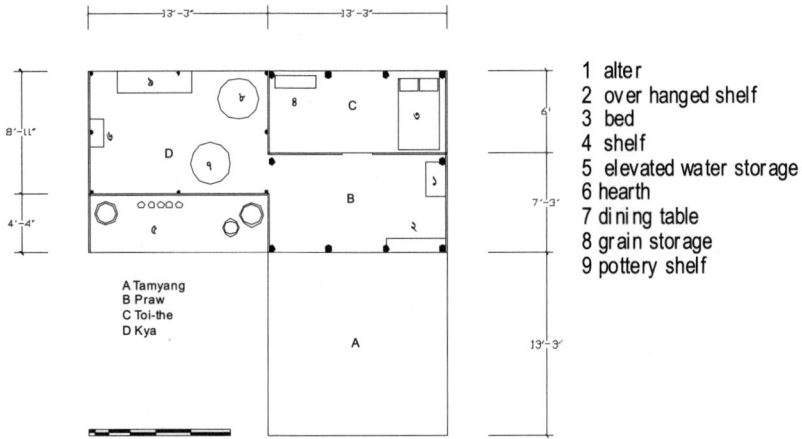

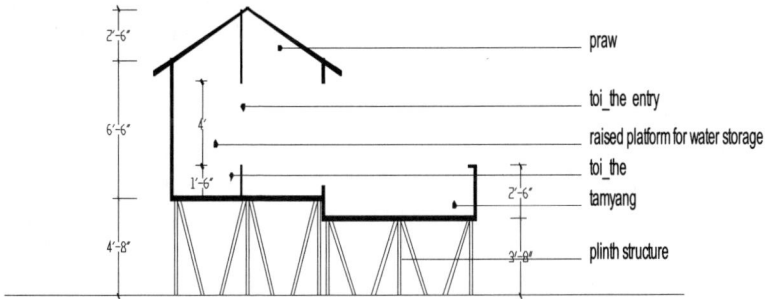

section through main room

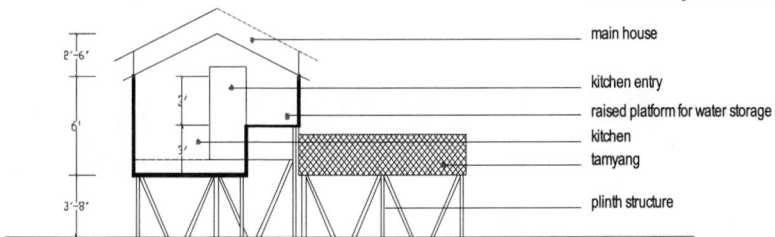

main house
kitchen entry
raised platform for water storage
kitchen
tamyang
plinth structure

section through kitchen

Mhrang Uing, Marma house showing internal layout.

One common feature is that families of all communities have a main house or room and at least one terrace. The main house, which is a large rectangle in shape, is usually divided into several rooms according to the user's needs. In a *Mhrang Uing*, the main house is divided into two rooms, *praw* (living room) and *Toi-the* (bedroom), by an internal wall. The floor level of the main house is raised 6" from the terrace and it has a separate overhanging *do-chala* roof. *Praw*, chiefly serves as a family living room and guest entertaining room. A sleeping arrangement is made if there's a son or guest staying overnight. Overhead compartments are used to store clothes, glasses, cigars, books, etc. and an altar carrying the statue of Buddha. *Toi-the* is the most private part of the house. To restrict entrance, the entry door to the bedroom is raised usually up to 1-6". The width of the room is limited to maximum 6'-0" corresponding to average human height.

The main room

The terrace is an elevated open or semi enclosed platform, which is attached to the front part of the main house facing the access road. It is reached either by climbing a notched tree ladder or bamboo built ladder. A wide range of outdoor activities like drying clothes, chilies, threshing corns, weaving, etc. are performed here. It is a space for relaxing and socializing.

Various uses of terrace

Raised water storage area and hearth

In most communities the kitchen is built separately with a separate roof system. In a Marma house, the floor height of the kitchen is lowered from 6" to 1' (150-300 mm) to segregate it from the main house. The kitchen also serves as a place for dining and food storage. After harvesting, a granary made of woven bamboo is placed in one corner to store the food grain. Over the hearth, an overhanging storage space is kept to dry gourds (water pots) and dry fishes. In Marma and Bawm houses, the kitchen includes a raised water storage area.

The underneath space of the raised house is used to store various things but chiefly for storing firewood and rearing animals i.e. pigs, ducks and hens. The hill people lead a very simple life. Hardly any furniture can be seen. Household tools and outdoor furniture are made using natural materials in a simple but innovative way.

Storage below the raised floor

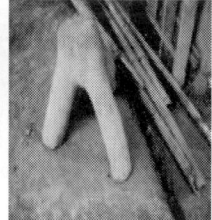

Building Materials

The houses are built using natural materials that are found in the forest. Timber, bamboo and wild grass are used in construction. Tree trunks (Garjan and *Gamari*) with branches 5"-6" in diameter are used as the main post and beam in stilt construction and wild grass is used for roofing. Bamboo is the chief building material in the CHT and used in various ways in different parts of house construction. Even to tie bamboo joints *ning* (bamboo strings) is used.

As a building material, bamboo has some advantages. It is relatively strong and stiff because of its hollow form and bamboo structures are good at withstanding storms and earthquakes. Bamboo houses are naturally ventilated and well lit as the wall panels and floor mat are made of weaved bamboo. Therefore windows are not seen very much in hill peoples' domestic structures.

The building materials require very low maintenance. Normally a house changes its thatched roof every three rainy seasons and the floor mat every ten years.

Construction Technique

Usually a house is built by the family and helped by the villagers in construction. The building technique and measurements are specific and this knowledge is preserved through generations. Even every structural element has different local terms in different hill communities. The construction process can be summarized into 3 main parts: floor, wall and roof construction.

Floor construction

The floor is constructed in a most innovative way through five successive steps. As the final floor rests on several layers of frames, the floor becomes more rigid and strong.

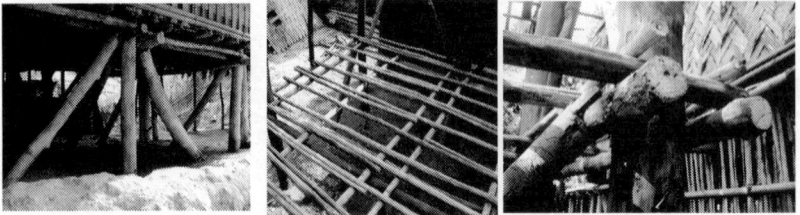

Floor construction

The following is a detailed description of a floor construction technique of the Marma community. At first the main posts (*Khong-the)* of the tree trunks are founded in the ground. Then pairs of tree logs 2.5" (70mm) in diameter are nailed to each row of posts in both parallel and perpendicular directions, which act as floor beams (*Kadat)*. Thirdly, round *Maday* bamboos are tied with bamboo lashes in between on the second layer of beams, which act as joists (*Pung-phi)*. Then round *Khayang* bamboo are added to attain a dense grid. These are the secondary joists (*Chang-bo)*. Finally, the floor (*Krang)* is placed, which is made of weaved bamboo.

Step-1

Step-2

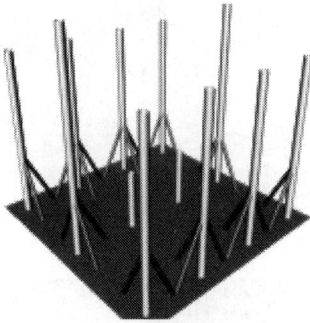

Step-3

Step-4

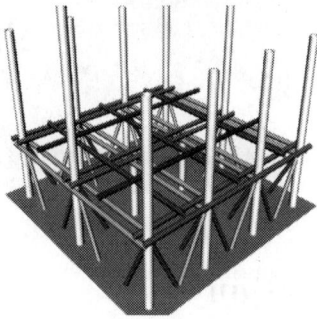
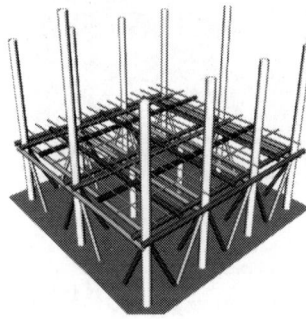

Step-5

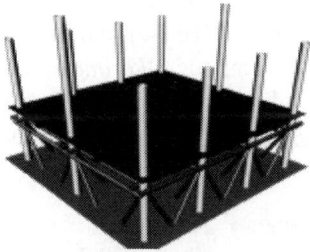

The successive stages of floor construction in a Marma house are shown in 3D, prepared by the authors.

Wall Construction

First a frame is constructed with the posts that rise from the ground and four bamboo posts (*Akang*) are placed diagonally in between them. Then the small bamboo or timber are tied or nailed in pairs horizontally and vertically forming a grid that is called *Tang* and *Akwat* respectively. Lastly, the wall (*Tharang*) made of woven bamboo is set.

Wall construction

Details:
A Kadat
(Post)
B Na-Prang
C Thow
(Main Beam)
D Wa-ow

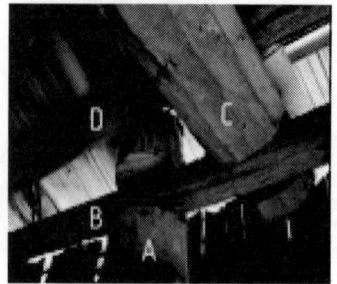

Roof Construction

In roof construction, a support (*Na-prang*) of tree logs is placed on the post to hold the main beam and the main beam (*Thow*) is placed on the support transversely to it. Then another support (*Wa-ow*) parallel to Na-prang is placed over the main beam. Round bamboos (*Akhrang*) are now placed at the direction of the roof slope and around it pairs of flattened bamboo (*Kwa-ow*) are tied together. Finally, the roof is covered with thatch (*Sakray*) or wild grass.

Roof Construction.

Houses sometimes have a very small window opening in the main room. Two types of windows can be found, pivoted and slide, while the main entry door is often hinged at the top.

Slide window   Pivoted window   Main entry door

The *machang* houses in Chittagong Hill Tracts gracefully stand strong. Vernacular architecture values the importance of social and cultural norms to individuals and society, and appreciates and ensures the coexistence of man and nature. Practitioners have a lot to learn from this indigenous architecture as it can provide guidelines for cost-effective and sustainable architecture. This can offer new vocabularies of domestic architecture appropriate for the CHT. Its traditional construction technique, if studied extensively, may open up new ways of using bamboo, which is still available and cheap as a building material.

Work Cited

1. Ahmed, K. I. and Kabir, K. H. (2005). *A Study of Traditional Housing of "Pahari" Communities in the Chittagong Hill Tracts (CHT)*. Dhaka, Bangladesh.

2. Brauns, C. and Löffler, L. G. (1990). *Mru, Hill People on the Border of Bangladesh*. Basel, Birkhäuser Verlag.

Traditional Foods of the Indigenous Peoples of the CHT

Ushing Prue, Nimaprue Marma, Ching Mo Sang,
and Lucky Chakma

The traditions, customs, rituals, and history of the 11 ethnic communities of the Chittagong Hill Tracks (CHT) are completely different from those of other people of Bangladesh. Their food habits and diversified foods have also their own distinctiveness. As they are nurtured by Mother Nature, a large proportion of their food also comes from nature meaning the forestland. The chilli, spices, and vegetables from their traditional *jum* cultivation are very different from those of the mainland in smell and taste. The kitchens of the hills are also very different. The main four sources of the delicious foods made by the hilly people are:

- *Jum* (source of mainly food grains, fruits, chilli, and spices)
- Forest (wildly grown vegetables)
- Market (dried fish, meat, and different kinds of easily available food)
- Homemade cakes (different indigenous people make different types of cake)

Jum

The entire indigenous communities in CHT were *jum* farmers or *jumias* from time immemorial. Though there have been many changes with the passage of time, the occupation of a large proportion of the Adivasis is still *jum* cultivation. This is the reason why their food habits, culture, and tradition—everything is dependent on *jum* cultivation. It can be said that their culture is the '*jum* culture'. The *jumias* select a convenient piece of land in the months of January and February. Then they cut all trees except the large ones. In the months of March and April, they burn the cut *jum* and plant mixed seeds of different crops like rice, cotton, melon, millet, beans, gourds, yam, maize, oil seeds, sesame, corn, pumpkin, ginger etc. in small holes. Different crops ripen at different times.

The followings are the major vegetables produced in *jum* and brief descriptions of the dishes the *jumias* prepare.

Tokhfol (*pungshi* in Marma, *amilay* in Chakma, *mokhongthay* in Tripura): Locally it's known as *tokhfol* because it tastes sour. It's red in color. The leaves of this tree are also edible. It's boiled with *nappi* or crab to make it taste less sour and more delicious. The plant bears flowers and fruits in the month of Kartik (October-November) and Agrahayan (November-December). Meat and dried fish can be cooked with the dried fruit.

Misti Kumra (pumpkin in English, *frungshi* in Marma, *shuguri gulo* in Chakma, *chak-ma* in Tripura): Pumpkins from *jum* are very tasty. Adivasis eat the flowers, leaves, and pumpkin shoots along with the fruits. The fruit is cooked with oil, crab or some other ingredients. The flower and shoot is cooked with *nappi* sauce and salt. Some people use ground wet rice to use in the pumpkin shoot curry. The pumpkin shoots are cut in little pieces and green chilli cut into slices are mixed into the boiling curry. When the curry is half boiled, ground rice is mixed into it. After some time it is removed from the fire. The stem of pumpkin is also cooked in various ways.

Chhoto tita begun (small bitter brinjal in English, *kojosi* in Marma, *begol bizi* in Chakma, *khongkha* in Tripura): It looks like brinjal but it is very small in size. Though it tastes bitter, it is very popular among the Adivasis. For cooking this, first it is divided into two pieces to bring out the seeds, which look like the brinjal seeds. The seeds come out when it is washed properly. Then green chilli paste is boiled in strong *nappi* soup. The vegetable is added when the fluid is heated and the dish is covered. After a while it is removed from the fire.

Morok pata (fowl leaf in English, *kramorok* in Marma, *talom pada* in Chakma, *Oraning* in Tripura): The size of the leaf is a bit large and it tastes a bit bitter. To reduce the bitterness, Adivasis use strong *nappi* in the curry. As the leaf is large, it is cut into small pieces to cook it with adequate salt in *nappi* curry.

Tita korola (bitter gourd in English, *gangkhashi* in Marma, *thide gulo* in Chakma, *Kanggla* in Tripura): Bitter gourd is a well known vegetable. Though the bitter gourd from *jum* looks the same, it is shorter and more round. This raw vegetable is finely chopped, mixed with salt, kept in the air for ten minutes, and washed. To reduce the bitterness strong *nappi* is added in the *nappi* soup. Green chilli paste and salt is added in the hot soup. When the soup is boiled, chopped bitter gourd is mixed with it to finish the cooking. The Adivasi people also eat its stem besides the vegetable. The procedure of the cooking is the same. However, to reduce the bitterness, it is cooked with pumpkin stem following the same recipe.

Chal kumra (ash gourd in English, *frunggraingsi* in Marma, *whomurow* in Chakma, *khakhulu* in Tripura). It is usually cooked with different types of meat like pork, deer meat and snail. The Adivasis eat the stem and shoots of Ash gourd. Ash gourd stem is boiled in water and is dried afterwards. A paste is made with dry chilli, *nappi*, ginger, and garlic and cooked with it. The shoot of Ash gourd is cooked in the same procedure like pumpkin shoots.

Orhor (*fiangkhungshee* in Marma, *domooch shumi* in Chakma, *khokkhelaing* in Tripura): It looks like bean. Adivasis usually cook tender *orhor*. Tender *orhor* is boiled and then the water is drained. One type of spicy mash is made with *orhor*, mixed with dried chilli smoked, *nappi* and ginger paste. It is also cooked with *nappi* curry. When it is ripe its seeds are taken out and soaked in water. It is cooked like *dal* by cleaning the seed's skin.

Chotto (small) *chinar* (*nyachaikyashi* in Marma, *sindire* in Chakma, *khaichaa* in Tripura): It looks like small pumpkin. It is eaten like a *bangi* fruit while unripe. Its taste is a bit sour so it is cooked with prawn, snail, and dried fish.

***Chinar* (*sueguecee* in Marma, *shindeyray* in Chakma, *thaisuemoo* in Tripura):** There are two kinds of *chinar* grown in *jum* (small and big). It is eaten like a plain land *bangi* (a kind of melon).

***Jum* aloo (potato in English, *faroai u* in Marma, *petti aloo* in Chakma, *thalairow* in Tripura):** It looks like a carrot. It is white in color. It is boiled with salt and eaten.

Shimul aloo (cassava in English, *mrolpieing* in Marma, *shimei* aloo in Chakma, *thabuchu* in Tripura): There are different types of this tuber shaped vegetable from *jum*. It looks like sweet potato. Indigenous people usually eat this during food scarcity. It is smoked or boiled with salt.

Kaon (Kaon rice in English, *cheh* in Marma, *houn chowl* in Chakma, *maishoy* in Tripura): It has small beads like sesame but its color is yellow. In different offerings, occasions, and festivals of indigenous people *kaon* is cooked as a sweet dish.

Bhutta (corn in English, *mokka* in Marma, *mokkay* in Chakma, *mycoonda* in Tripura): A well-known food grain, the corn of *jum* is soft and delicious. Corn is smoked or boiled. Dried corn is also fried.

Dherosh (okra or lady's finger in English, *banggyosi* in Marma, *dagashumi* in Chakma, *moiromee* in Tripura): A common vegetable grown in *jum,* it is popular among the *jumias.* Especially those who stay in the *jum* for a long period like to eat the boiled okra with chilli paste. It is also cooked with *nappi* and oil.

Begoon (egg plant or brinjal in English, *khayriengsee* in Marma, *begoon* in Chakma, *fathao* in Tripura): There are two or three kinds of egg plant in the *jum—jati* begoon, *tita* (bitter) begoon, and *kata* (thorn) egg plant.

Marfa (*noprosi* in Marma, *mar-ma* in Chakma, *mathai* in Tripura): It looks like cucumber. Grown in *jum,* it is a common and favorite vegetable for the indigenous people.

(top row from left) *Theipi*, a wild fig and hill chilli. (second row from left) *Gach Owl*, a special kind of mushroom and bamboo shoots. © Philip Gain

Moreech (chilli in English, *norosi* in Marma, *mooreech* in Chakma, *thaisueoo* in Tripura): Chili is very familiar in our country. Chili is used for any kind of spicy cooking.

Chichinga (snake gourd in English *thoyethasi* in Marma, *hoidey* in Chakma, *chingga* in Tripura): A common vegetable the taste *jum,* the snake gourd grown in *jum* is different.

Lau (bottle gourd in English, *buh si* in Marma, *hudugulo* in Chakma, *moilau* in Tripura): There are two types of bottle gourd grown in *jum.* One is only for eating and the other, when mature is dried and emptied for use as a water bottle.

Borboti (long beans in English, *oway payosi* in Marma, *shumi* in Chakma, *subai* in Tripura): The look of *jum* long beans is the same as those seen in the plains land but its color is dark purple.

***Jum akh (Akh* is sugarcane in English, *prong* in Marma, *jaytheyna* in Chakma, *masingga* in Tripura):** This plant looks like sugarcane and it is eaten like sugarcane.

***Jum* khoi rice (*jum* puffed rice in English, *prongchi* in Marma, *jaytheyna* in Chakma, *nangabrung* in Tripura):** This variety of paddy is grown in *jum* mainly for making puffed rice. It is sold also sold for a good price. At present very few *jumia* people grow it, as the land has lost its fertility.

Delicious local dishes served in a forest village in Bandarban.   © Philip Gain

Vegetables from the Forest

Kolar thor or mocha (banana bud in English, *nopyopho* in Marma, *holathtur* in Chakma, *mungkhukthuh* in Tripura): The hill people cook it in different ways. Some of them cook it with small amount of *nappi* mixed with salt and boil it and use chilli paste. It is also cooked with *nappi* and a few egg plants. Then it is also boiled and fried with oil.

Kochu shak (arum leaf in English, *pringbungshey* in Marma, *hujushak* in Chakma, *thakang* in Tripura): This is one of the desired vegetables of the hill indigenous people. It is boiled and cooked with lot of *nappy*, oil, spices, and chilli. Sometimes they use a bit of sour to reduce the itching of the throat.

Gimmi shak (*shak* is leaf vegetable in English, *gaingga* in Marma, *dhimetidea* in Chakma, *bokkhathoi* in Tripura): This vegetable is very common to Adivasis. Usually it is grown near the river. As it is so bitter, it is cooked with a lot of *nappi* and chilli.

Minjiri shak (*moijheri* in Marma, *minjiri shak* in Chakma, *mozli* in Tripura): It looks like a tamarind leaf and tastes bitter. The size of the leaf is bigger than that of tamarind. It is boiled and cooked with dried prawn and *nappi*.

Shimul fool (silk cotton flower in English, *lapaiboaung* in Marma, *shmayeefool* in Chakma, *bauchungba* in Tripura): This flower makes a very popular dish. The petals are thrown off and the stem is dried in the sun properly. It is cooked with both *nappi* and oil.

Tara gach (star tree in English, *chingyang* in Marma, *tara* in Chakma, *mueetangsue* in Tripura): Its tender part is cooked. It is boiled and cooked with chilli, *nappi,* or oil.

Baash korol (bamboo shoots in English, *hmoy* in Marma, *bhacchuri* in Chakma, moiyaa in Tripura): It is boiled and dried; again before cooking it is boiled. Tender parts are cooked with oil or *nappi*.

Olkochu (*pringfi* in Marma, *woolkuchu* in Chakma, *bataymatha* in Tripura): There are different kinds of *olkochu* in the forest. Some are cooked with oil and others with *nappi*.

***Trice shak* (*choroitok* in Marma, *hatholedhingi* in Chakma, *katha muchoo* in Tripura):** Cooked with *nappi* or oil. It is cut into small pieces and fried.

***Khona* (*krongsasi* in Marma, *khonagulo* in Chakma, *taukharung* in

Tripura): *Khona* is a medicinal plant. This plant's tender leaf and fruit is eaten. *Krong* means cat and *sha* means tongue in Marma language. It means cat's tongue fruit. This is a little bit long in shape. It is burnt on the fire, and the skin is peeled. It is cooked with oil and *nappi* or separately.

Bonoj moshla gach (wild spices tree in English, *falagong* in Marma, *palachangaee* in Chakma, *bring* in Tripura): Its fruit and roots are eaten. It is not cooked as curry. It is used to bring flavor in the curry.

Beter aga (cane shoots in English, *kring* in Marma, *baythagi* in Chakma, and *rynengfang* in Tripura): It is a favorite vegetable to the Chakmas and Tanchangyas. They call it *bedagi* or *beragi*. The tender cane is boiled and cooked with chilli and *nappi*.

Banger chhata (mushroom in English, *hmo* in Marma, *ul* in Chakma, *muikhung* in Tripura): There are four to five kinds of mushrooms in the forest. Most of the indigenous peoples eat these mushrooms. It is cooked with oil or *nappi*.

Bonoj alu (wild tuber/potato in English, *pringfie* in Marma, *jarbo* potato in Chakma, and *thum* in Tripura): These tubers/potatoes are found in the jungle. The Adivasis have been consuming these for generations. Those are *mon alu* (in Chakma), *hoang alu, Dua alu, Kirija alu*, etc. The Adivasis also search for these tubers during a food scarcity. They cook these tubers/potatoes as per their own choice. Sometimes the potatoes are cooked with *nappi* or oil.

***Pathar kochu* (taro in English, *kyakpring* in Marma, *sheelkochu* in Chakma, *dalaimoitto* in Tripura):** The stem of this taro is edible, specially the tender ones. It looks like grass. This type of taro is not cooked with oil. Most of the taro is cooked with *nappi*. Plenty of chillies make it tasty. It is found near the rivers (where there are pebbles).

Bonoj lata shak (wild creeper *spinach in English*, *orongkhyoing* in Marma, *mormosha amile* in Chakma, *mukhoeepoipo* in Tripura): These leaves are light green. This *shak* is cooked with crab, which is found below the hill where a stream flows. The indigenous people use this *shak* also as a medicinal plant. It is usually cooked with *nappi*.

Star type aromatic tree (*chingyang* in Marma, *chengyei* in Chakma, *bring* in Tripura): There are three kinds of wild aromatic trees and they are different in aroma. They are usually found in moist places. The fruit is cut into pieces and cooked with oil or *nappi*. The indigenous people like to eat it by boiling the star tree fruit adding *nappi* water and salt and then served with chilli sauce.

Bandar marfa (*changgrisi* **in Marma,** *bandarmarma* **in Chakma,** *mathichakrong* **in Tripura):** It is small in size and its body is covered with thorns. At the tender stage it is bitter but sour when it is ripe. It is generally cooked with *nappi*. Some people cook it with oil. Most of the Adivasis like to eat it as sauce.

Bonoj shak (wild spinach in English, *fauma* **in Marma,** *ambooj shak* **in Chakma,** *mueloro* **in Tripura):** Though this *shak* is wild it is found mainly in *jum*. Its leaf is thrown away and the tender part is taken. Then it is cooked with *nappi* and egg plant.

Pudina (mint in English, *hungfowee* **in Marma,** *ozhon shak* **in Chakma,** *osoondoi* **in Tripura):** It is a type of vegetable, which gives a twinge sensation in the tongue. Because this vegetable gives a sensation, some people boil it before cooking. It is cooked with necessary ingredients. It is also fried.

One type of tender leaf of the forest tree (*wocri wocry* **in Marma,** *bathbattay shak* **in Chakma,** *muekhani* **in Tripura):** This leaf is found from a type of tree in the deep forest. This leaf is eaten raw with *nappi* and chilli paste.

(left) Varieties of vegetables from the hills in a local market. (top right) A hill woman at her kitchen. (bottom right) *Tok pata* (sour leaf).   © Philip Gain

Aam pata (Mango leaf in English, *saranu* in Marma, *amshak* in Chakma, *thaichusuek* in Tripura): Though this leaf is cooked yet majority of the people eat raw mango leaf with *nappi* and chilli paste with rice.

Guthguttya shak (sueduee neowai in Marma, *guthguttya shak* in Chakma, *thaishre* in Tripura):* This *shak* is washed and eaten raw with chilli paste.

Tetofol (*bitter fruit* in English, *tohkhasi* in Marma, *tetholgulo* in Chakma, *doshkay* in Tripura): This fruit is cooked as curry. Because it is too bitter, it is cooked with an increased amount of *nappi* and rice flour. Some mix a little bit of oil into it.

Jongli kathal (wild jackfruit in English, *tohpenasi* in Marma, *chamani kattol* in Chakma, *jrang* in Tripura): It is eaten like jackfruit.

Bon potol (wild parwal in English, *sofoeye see* in Marma, *jharbooa porol* in Chakma, *fro* in Tripura): It is boiled to eat.

Shikakai (acacia concinna in English, *kangboinu* in Marma, *kojoi* in Chakma, *mokhoicheechai* in Tripura):* It is one of the Ayurvedic Medicinal plants. Adivasis use it as *shak.* It is used in chicken soup. It tastes sour.

Chal kumra, a popular vegetable in the hills.   © Philip Gain

Foods from Market

The meat of the hunted animals and birds are available in the hill market. The sources of the fish sold in the market are the river, fountains, and the Kaptai Lake. The following are descriptions of fish and meat available in the market:

Jhiri kakra (stream crab in English, *mrongcanai* in Marma, *monkangara* in Chakma, *toishakhanggri* in Tripura): This crab is found in the deep forest near small streams under a stone or in a hole. It is caught at night while it comes out in search of food. The crab is cooked with oil or *nappi*.

***Guishap* (monitor lizard in English, *foyesa* in Marma, *guee* in Chakma, *mohfu* in Tripura):** The meat of *guishap* is mixed with onion, garlic, ginger paste, rice half ground, salt and then cooked with oil. Green turmeric leaf is added before removing it from the heat. This adds an aromatic smell to the curry.

Choto octopus (small octopus in English, *rugra* in Marma, octopus in Chakma, *leegrow* in Tripura): It is cooked with small amount of water and salt along with onion, garlic, ginger, chilli paste and turmeric.

(top row from left) *Nappi* or *sidol*, an essential curry paste in the hill peoples' kitchen made from fish or shrimp and *jhiri kakra*. (second row from left) frog, a popular food in the hills and crab from the plains.   © Philip Gain

Jhee jhee poka (cricket in English, *proee* in Marma, *gumurow* in Chakma, *khrangbu* in Tripura): Mixed with turmeric and salt it is fried in oil.

Shamuk (snail in English, *khuruk* in Marma, shamuk in Chakma, *scumboo* in Tripura): It is cooked with *nappi* mixed with water, chilli paste, and sour leaf.

Jhinuk (oyster in English, *goong* in Marma, *celon* in Chakma, *goong* in Tripura): It is mixed with turmeric and salt and then fried in pork oil.

Bangachi (tadpole in English, *folongsa* in Marma, *begana* in Chakma, *jaypro* in Tripura): Tadpole is mixed with ginger, chilli, onion, garlic paste, and then stir fried.

Shona bang (frog in English, *posowah* in Marma, *bhawba bang* in Chakma, *koahbaing* in Tripura): It is mixed with ginger, chilly, onion, garlic paste and then stir fried.

Shomudro *pathar kuchi* (sea *pathar kuchi* in English, *kyakro* in Marma, *pathar kuchi* in Chakma, *toweekhamai* in Tripura): It is washed thoroughly and then mixed with dry chilli, salt and tamarind paste.

Shutki (dried fish) found in local markets in abundance is the main source of protein for people who live in the CHT.   © Philip Gain

Hangor shutki (dried shark in English, *nyamai khro* in Marma, *angor mach* in Chakma, *namai* in Tripura): It is mixed with chilli, ginger, onion paste, and salt.

Chapila **shutki (*chapila* dried fish, *chapila khro* in Marma, *chapila* in Chakma, *nakhekha* in Tripura):** *Chapila* dried fish is used in white pumpkin to make a delicious and tasty dish.

Pata shutki (leaf dried fish in English, *fakshay* in Marma, *pada mach* in Chakma, *likayotowlia* in Tripura): Leaf dried fish is smoked and mixed with green chilli and salt.

Chhuri shutki (chhuri **dried fish, *neosoroye* in Marma, *suri shugunee* in Chakma, *nasaroway* in Tripura):** White pumpkin vegetable is mixed with turmeric, chilli powder, ginger, onion and garlic paste. *Chhuri* dried fish is washed and mixed into it before five minutes of cooking. After five minutes the curry is cooked properly.

Baila shutki (baila **dried fish, *senewaisa* in Marma, *baila shugunee* in Chakma, *chicheeree* in Tripura):** It is smoked and served with rice. It is mixed with different curries.

Loyetta shutki (loyetta **dried fish, *nyodumoung* in Marma, *loyetta shugunee* in Chakma, *nadamoung* in Tripura):** It is cooked with *nappi* mixed with water and shredded potato.

Chingri shutki (dried shrimp in English, *mojowai kroh* in Marma, *ejaguri* in Chakma, *authukrung* in Tripura): Dried shrimp is used in different curries to make them delicious.

Shukurer mangsho (pork meat in English, *woksa* in Marma, *suegorara* in Chakma, *wokhung* in Tripura): Pork meat is mixed with onion, garlic, ginger, green chilli and turmeric paste and cooked with salt and a moderate amount of water.

Comment: The fish, meat, and vegetables available in the market are prepared as daily food by the indigenous communities. These dishes bear traditions of their life style.

Pitha (Home Made Cakes)

Varieties of *pitha* are delicacies to indigenous peoples particularly at the time of festivals and *pujas* (worships). Different kinds of food and *pitha* are prepared during these festivals. When relatives and guests visit their homes, they serve these homemade delicacies. In the hill tracts region individual families among all indigenous communities prepare *pithas* for *Baishabi,* a major festival. The Marma young boys and girls divide into groups and make *pithas* late at night for *Probarona Purnima* and *Baishabi* festival (probarona is fulfillment of wish and purnima full moon). *Pithas* prepared all through the night at a festive mood are distributed from house to house in the morning. Popular *pithas* prepared by different indigenous peoples are briefly described below:

Marma: *Pitha* is *muhng* in Marma language. They prepare varieties of *pitha* at different *pujas* or occasions. Among these *fisha muhng* is a favourite *one.* *Binni* or normal rice powder is mixed with a small amount of water to make it soft. The soft mixed rice powder is made thin round shaped by hand, and coconut and sugar stuffed into it, then half shaped and edged closed with finger tips. It is wrapped with a banana leaf. When *cheh tobong muhng–binni* rice is moist and soft it is ground. Then it changes to flour and sieved. The flour is mixed with coconut and sugar, which makes a special pattern. As it makes a special pattern so in Marma language it is called *cheh tobong muhng.* It is like Bangali's *pati shapta.* Others are *gung muhng (shell pitha), musho (mou pitha), kayang chi gyaw muhng (bini pitha* oil fry), *khooa pong muhng* (steamed *pitha*), *mundi* (rice noodles), *ro choo muhng (tip pitha),* etc.

Lushai: The Lushai, all Christians, do not prepare their traditional *pithas.* But they prepare a *pitha* (*zawjiang*) for different religious occasions. This *pitha* is prepared with *binni* rice flour mixed with banana or without banana wrapped with banana leaf and heated.

Tripura: The Tripuras prepare different kinds of *pitha* for various occasions. *Maimi pungau* is the favourite *pitha* of the Tripuras. *Binni* rice flour is mixed with water and wrapped with banana leaf. Then it is boiled in a special way by heating it in a pot. In Tripura language steamed *pitha* is called *maimi pung aumong.* One kind of *pitha* is made by *binni* rice flour mixed with water and made round shaped and fried with pork oil. It is called *owakhfah awah pitha. Maimi wasung pitha,* another *pitha* is made with binni rice flour. The rice flour is filled in a bamboo hollow tube. Then for baking the hollow tube is burnt in the fire. *Binni* rice is also cooked by soaking it in water and then wrapping it in banana leaf in a special way by heat. In Tripura language this *binni* rice is called *awah bhangguho.*

Tanchangya: The Tanchangya people prepare different kinds of *pitha* for festivals and family occasions. *Guri pitha, kolagula taley pitha, binni chalar* *pitha, bhapa pitha, and kolagula taley pitha* are some of popular *pithas*. To prepare *kolagula taley pitha*, rice powder or *binni* rice powder is mixed with banana. Then the mixture is fried in hot oil by small amount. To prepare Molasses *pitha*, rice powder is mixed with molasses and coconut cut into small pieces and wrapped in banana leaf or jackfruit leaf. Then in a special way it is dried by heat.

Khumi: Khumis do not make any special *pitha*. What they prepare is *tuho ko*. *Binni* rice is soaked in water and mixed with coconut pieces. Then it is wrapped in a banana leaf and heated. Another type of *pitha* is made by *binni* rice filled in a bamboo hollow tube and then burnt in the fire. They call this *pitha bipaho torana*. In Khumi language *binni* rice is called *shangti biho pa*.

Bawm: The Bawms do not have varieties of *pithas*. *Binni* rice is soaked and filled in a hole below a dry bottle gourd. Then a pot is filled with water and it is heated. The *binni* rice filled in dry bottle gourd is kept on top of the pot. By the steaming method the *binni* rice is cooked. In Bawm language it is called *thaubuho hoot*. The Bawms make another kind of *pitha* named *thaubuho chawth*. On different occasions banana is dried and served. Dried banana is called *banla char*. Sweet rice is made by *kaon* rice cultivated in the *jum*. In Bawm language it is called *booton chawth*.

Khyang: *Pitha* is prepared with *binni* rice flour with a mixture of sugar, coconut, and then fried in pork oil. Another kind of banana *pitha* is made by *binni* rice flour with a mixture of sugar in water as per requirement and wrapped in banana leaf and heated.

Chak: The Chaks make *pitha* with *binni* rice. In different *pujas* or occasions they prepare sweet rice with *kawon* rice and mix with sugar, cow milk and nuts. They also make small tip *pithas* with rice flour.

Mru: The Mru people prepare one kind of *pitha* called *stoyrong* with *binni* rice soaked for half an hour and then filled in raw and tender bamboo hollow tube and burned in the fire. *Binni* rice or scented rice cultivated in *jum* is soaked for a night and mixed with pork oil and filled in raw bamboo hollow tube and burned in fire. In Mru language it is called *pak shao komrung pitha*.

Pankhua: The Pankhuas do not have the custom of making *pithas*. In different occasions they soak *binni* rice kept in hollow tube and then cooked like Bawms. *Pitha* is also made by cooked *binni* rice fried with pork oil. The *binni* rice is made powder by a husking pedal and mixed with sugar, coconut, and water. The mixture is dropped by tablespoon in high heat pork oil and dried. They call it *taylay pitha*.

Chakma: The Chakmas prepare varieties of *pitha* (*Peedhay* in Chakma language) in different festivals and occasions. The famous *pithas* are *baura peedhay, shannay peedhay, bang peedhay, binni peedhay or hoga peedhay, kur angul peedhay. Baura peedhay*—*binni* rice powder is mixed with water, chopped coconut, and nuts and then fried in oil. In the past pork oil was used but at present soybean oil and mustard oil is used. *Shannay peedhay*—rice is made powder by a husking pedal, round shaped, and put into hot water. The half boiled rice powder is mixed and round shaped *pitha* is made. The half boiled rice powder is mixed with coconut, nuts, and molasses. *Binni peedhay* or *hoga peedhay*—*binni* rice is made powder by a husking pedal, mixed with water thoroughly, and wrapped in a jungle leaf. Then the *pithas* are steamed in a hot water pot. *Bang peedhay*—rice powder is mixed with the ingredients as required and made long size and wrapped in banana leaf. Then it is steamed as *hoga peedhay*.

Special thanks to: Keywshaw Prue, Lalchhani Lushai, Sattaha Panji Tripura, Rozina Tripura, Buddyojyoti Chakma, Wong Ching Marma, and Simon Amlai.

A Bangali family having the taste for local foods in a Chak village in Baishari, Bandarban.   © Philip Gain

Militarization of the CHT and Its Impact on the People

Jenneke Arens

Roots of the CHT Conflict

Significant interference in the resource-rich hilly area started with the annexation by the British colonial rulers in 1860 to safeguard their political and economic interests. The British named it the Chittagong Hill Tracts (CHT) and established indirect rule. The indigenous *rajas* (chiefs) were appointed as collectors of revenue under the supervision of a Superintendent and later of the Deputy Commissioner (DC). All land in the CHT was declared government property and the indigenous peoples were given tenancy rights over the land, although according to indigenous customary rights land is communally owned. The British also introduced plough cultivation and the transformation of a self-reliant economy based on *jum* (shifting or swidden) cultivation to a market-dependent economy was set in. The underlying aim was to increase the land revenue and make it easier to collect revenue from the people once they were settled in one place. The government would then be less dependent on the chiefs for tax collection. Commercial and urban centres grew and commodity production and circulation of money increased. Bangali immigrants took most advantage of the introduction of plough cultivation as they were used to plough cultivation in the plains. The introduction of plough cultivation also led to social and economic differentiation among the indigenous peoples. Those in the river valleys, mostly the Chakma and Marma, managed to acquire plough land and benefited mostly, whereas groups on the mountain ridges, such as the Mru, Bawm and Pangkhua, continued to live mainly from *jum* cultivation.[1] With the argument to curb *jum* cultivation the British declared one-fourth of the area as forest reserved and access to these reserves became strictly limited, thus depriving the indigenous peoples from a large part of their collective *jum* land. As a result the pressure on land increased and the fallow period of the *jum* cycle decreased from 15-20 years to 8-10 years periods.[2]

In 1900 the British introduced the so-called '1900 Regulation'. This provided the area with a special administrative status and restricted further settlement of Bangali plainspeople in the hills. With the 1900 Regulation also a special judicial system was introduced to the CHT and the transfer of indigenous peoples' land to Bangalis from the plains was prohibited. All land transfers needed the approval of the DC in consultation with the local headman. The introduction of the 1900 Regulation was obviously for strategic reasons: the indigenous people who had strongly resisted invasion by the British were to be pacified to serve as a buffer against other 'wild races' on the frontiers of the British empire. However, it did give the area a special status, recognised the specific identity and culture of the indigenous peoples and protected their (landed) interests.

After independence in 1947 the successive governments of East Pakistan and later Bangladesh, failed to acknowledge the deteriorating situation and ignored the legitimate demands of the indigenous peoples to retain the protection of their specific identity. The military regime of Ayub Khan abolished the special status of the CHT in 1964. The result was an increasing influx of Bangalis from the plains and a growing exploitation of the indigenous population.

After independence of Bangladesh Sheikh Mujibur Rahman rejected the indigenous peoples' demand for regional autonomy, retention of the 1900 Regulation and a ban on influx of Bangalis. Although Sheikh Mujib himself had led the Bangali people in the struggle for their own Bangali identity and culture, he failed to recognise a similar demand of the indigenous peoples as legitimate. He told them to forget their ethnic identities and be 'Bangalis'. He also threatened to flood the area with Bangalis and military troops if the hill people would stick to their demands. Following Mujib's denial the indigenous peoples formed the *Parbatya Chattagram Jana Samhati Samiti* (PCJSS) in 1972, and a year later it's armed wing, the *Shanti Bahini* (peace force). The PCJSS introduced the term *Jumma* as a collective name for the 11 different ethnic groups, referring to the traditional *jum* cultivation, which is an important common component of their culture and identity.

General Ziaur Rahman, who came to power in 1975 through a military coup, ordered full militarization of the CHT and settlement of large numbers of landless Bangalis from the plains in the CHT. In reaction to this the *Shanti Bahini* carried out its first armed attack on a military outpost in 1976. When the settlement of Bangalis increased the *Shanti Bahini* started carrying out attacks against settlers as well.

General Ershad who grabbed power in 1982 through yet another military coup pursued the same counter-insurgency policies. The CHT had become an area under military occupation and the security forces in the name of counter-insurgency against the Shanti Bahini perpetrated massive human rights violations. Levene (1999:346) argues that a 'creeping' genocide of the *Jumma* peoples in the CHT has been taking place and the refusal of the *Jumma* people to be passive victims made him remark:

> The roots of genocide in the CHT do rest *in part* in the refusal of the *jumma* either to lie down and die quietly, or alternatively, to accept a place within the Bangladeshi scheme of things, for instance as colourful but otherwise harmless exotica, weaving carpets and dancing for the tourists, in some ethnographic zoo. Instead, their tenacious and bloody fight-back against state and settler encroachment alike, and their articulation of their political right to self-determination, has challenged the very notion of a religiously and culturally unified Bangladesh (Levene 1999:363).

Reports about human rights violations by the security forces started trickling out and General Ershad was under increasing international pressure from donor governments to come to a settlement with the *Jumma* resistance. Negotiations between the PCJSS and the respective governments have taken place from 1985 till 1989 without coming to any agreement. In December 1990 General Ershad was ousted by a mass movement that ended almost 15 years of military rule and parliamentary democracy was restored. In 1992 negotiations between the PCJSS and the subsequent governments resumed, but it was not until 2 December 1997 that a Peace Accord was signed between the PCJSS and the government, then headed by Sheikh Hasina of the Awami League. The opposition parties led by the Bangladesh Nationalist Party (BNP, the party founded by General Ziaur Rahman in 1978 to legitimise his military rule) and the Bangali settlers agitated strongly against the accord and rejected it as a sell-out. On totally different grounds a section of the *Jumma* people criticised the Peace Accord. They had been active in mass organisations such as the Hill Students' Council, Hill Peoples' Council and Hill Women

Federation, operating in close cooperation with the underground PCJSS. Their main criticism was that the main demands of the PCJSS--regional autonomy, constitutional recognition of the identity of the *Jumma* people and withdrawal of the settlers--had not been fulfilled. They declared that they would continue the struggle for full regional autonomy by democratic means and in December 1998 they formed the United Peoples Democratic Front (UPDF). Shantu Larma, leader of the PCJSS, took the opposition against the accord as a serious offence and a violent internal feud between the two parties was the result. This feud is continuing till today.

In 2010 the UN Permanent Forum on Indigenous Issues (UNPFII) requested Lars-Anders Baer, a former member of the UNPFII and a member of the CHT Commission[3], to assess the implementation of the Peace Accord. In his report to the UNPFII the 'Status of Implementation of the Chittagong Hill Tracts Accord of 1997' Baer described how the army abused its power in the CHT. One of his recommendations to the UN was to strictly monitor and screen the human rights records of army personnel before recruiting them for UN peacekeeping operations (Chakma 2011). Since 1988 the Bangladesh army has been involved in peacekeeping missions and Bangladesh is one of the biggest contributors of personnel to the UN peacekeeping forces. Peacekeeping missions are a big source of income for army personnel and for the army as a whole. Apparently Baer's recommendation worried the army and in a move to counter Baer's report, the DGFI (Directorate General of Forces Intelligence, the intelligence wing of the army) directed the government not to use the term 'indigenous' anymore (IWGIA report 14, 2012). Subsequently the government refused to officially recognize the indigenous peoples as "indigenous" in its 15th amendment to the Constitution despite a strong combined movement of the indigenous peoples from the CHT and the plains for recognition. The 15th amendment that was passed by the Parliament in June 2011 refers to the indigenous peoples as "tribes" or "small ethnic communities". This shows the control that the army has over the government.

The Peace Accord has been signed 15 years ago, but it still remains largely unimplemented, although Sheikh Hasina returned to power in 2008 and one of the election promises of the Awami League had been to fully implement the Peace Accord. The area remains under heavy military occupation despite a clause in the Peace Accord that the army will be withdrawn to the six existing cantonments in the CHT.[4] The army remains the de-facto ruler in the CHT, Bangalis continue to settle and serious human rights violations are still taking place.

Militarization

Under the regime of General Ziaur Rahman the 24th Infantry Division Chittagong was set up and the GOC (General Officer Commanding) was made in charge of operations in the CHT; the hills were heavily militarised. Apart from the already existing cantonments in the three district capitals another three cantonments were constructed in Dighinala, Ruma and Alikadam and gradually the number of military camps all over the CHT increased. General Ershad continued and intensified the military occupation of the CHT. Sixty percent of the Bangladesh army was deployed in the CHT (Jérémie Codron, 2007:[60]).[5] That means some 70,000 army troops on an indigenous population of 500,000 at that time. On top of that there were troops of the BGB (Bangladesh Border Guards, previously Bangladesh Rifles- BDR), Ansar (a paramilitary force) and the Village Defence Party (armed Bangali civilians). In total there were some 550 camps of the security forces during the insurgency period. By 1985 the defence budget had increased by more than 400%, compared to 1973. To put this figure in perspective, over the same period the budget for health had risen only by 18% and by 1985 the defence budget was more than 3 times higher than the health budget, whereas in 1973 it had been lower.[6] It is common knowledge that unofficially military expenditure is even much higher than is indicated in the government budget. A lot of the military expenses are budgeted under various other categories, such as telecommunication and housing, thus disguising the military purposes. It is also common knowledge that former President Ershad assured himself of the loyalty of high army officers by allowing them unbridled corruption and loans against very low interest rates. Large amounts of money from foreign aid disappeared in the pockets of military officers. Ershad himself had a good share of the cake as well. After his downfall and arrest in December 1990 newspapers reported that local currency notes equivalent to a total value of about $1 million were found cash in his house (Arens 1997:59). Besides, army personnel were involved in lucrative illegal logging in the CHT with the cooperation of the forest department, looting a large part of the forest resources. Hardly any of the original forest in the CHT is left due to logging while the authorities unjustly blame the indigenous peoples' *jum* cultivation for this.

The army carried out large-scale counter-insurgency operations in indigenous villages, claiming that the civilian population were providing food and shelter to the *Shanti Bahini*. In such operations the army often burnt down indigenous peoples' houses, sometimes even whole villages to force the reluctant people to move to cluster villages (I will elaborate this further on). Needless to say that the indigenous people were frightened whenever they met soldiers on the road or saw them approach their village. All kinds of restrictions were also imposed on the civilian population, such as limiting the amount of rice and clothes they could buy to prevent them from supplying the *Shanti Bahini*. On their way to the market or elsewhere people had to pass through army check posts where they were searched, intimidated and harassed.

As part of the counter-insurgency programme, the army has also been carrying out so-called ëpacificationí programmes. In sharp contrast to the military operations these were civilian programmes designed to "win the hearts and minds" of the civilian population by so-called 'friendship programmes', such as small-scale income generating projects, construction of schools, temples and roads and providing health care. Even after the Peace Accord the army is still involved in 'pacification' programmes in the CHT, thus weakening the civil administration. These are programmes that in a democratic society should be carried out by the civilian government.

In fact the army still controls the CHT and is very reluctant to withdraw its camps. One reason is that the CHT serves as a training ground for army personnel to be deployed in UN Peace Keeping Operations. Peacekeeping missions are a lucrative source of income for army men and for the army as a whole. According to finance minister A.M.A. Muhith in his budget speech in June 2012, 11,000 Bangladeshis are engaged in UN peace keeping missions and over the last three years peace keeping missions have earned in total Tk.4,430 *crore* (more than half billion US Dollar) in foreign exchange.[7] Part of the money earned by the army in peace keeping operations is said to be invested in the *Sena Kalyan Sangstha* (SKS), a big enterprise owned by the army (IWGIA report 14, 2012:22). The SKS has large investments in land and real estate and is involved in big business, such as the production of goods varying from cement, televisions and light bulbs to flour and ice cream. Besides, army personnel have other great benefits from serving in the CHT. Apart from the power and extra financial benefits it is common knowledge that army personnel in the CHT run profitable businesses. Most of the forest has been looted by army personnel and (ex) army men have leased or bought vast tracts of land in the CHT. To give some examples, General Matin, former GOC, has a lease of 275 acres of land in Bandarban District and allegedly sold 75 acres of the leased land to a relative, violating the terms of the lease contract; Colonel

Oli Ahmed (retired), former Minister of Communication, has leased 150 acres of indigenous people's *jum* land and Lt. General (retired) Harun-Ar-Rashid and his company Destiny 2000 grabbed more than 5,000 acres of land from more than 1,000 indigenous and Bangali families for commercial tree planting in Bandarban District (IWGIA report 14, 2012:22). Lt. Gen. (retired) Harun-Ar-Rashid and Destiny 2000 came in the news in April 2012 as his company was allegedly involved in illegal banking and other irregularities.[8] According to a recent report by Kapaeeng Foundation (2012) Major (retired) Mostafa Zaman control hundreds of acres of land in Bandarban District. It was also reported that, according to the local *upazila* chairman and the local headman, Major Mostafa Zaman had collected a permanent resident certificate for the CHT by forgery and had filed false cases against local people.

In order to prove that the presence of the army is still indispensable in the CHT the army allegedly patronises armed groups such as the *Borkha* (Veil) Party, a group similar to the *Mukosh Bahini* (Mask Forces), which was involved in attacks against *Jumma* activist organisations in the 1990's. For the *Borkha* Party unscrupulous and corrupt indigenous men are recruited; some of them have deflected from the JSS or UPDF and are allegedly used by the army to carry out armed attacks both on the JSS and the UPDF, also to feed distrust between these two parties, which sometimes accuse each other of these attacks.

Bangalisation

An important part of the counter-insurgency programme was the settlement of more than 450,000 landless Bangalis from the plains in the CHT between 1979 and 1983 under a secret government transmigration programme. During Ziaur Rahman's regime plans to settle hundreds of thousands of landless Bangalis from the plains in the CHT were developed in secret and from 1978 the settlement schemes started and ran parallel to the militarization programme. In a secret memorandum dated 15 September 1980 from the Deputy Commissioner of the CHT to government officials in other districts guidelines were given for the second phase of the settlement of landless non-tribal families from the plains. Each family would be given five acres hilly land, four acres of mixed land or two and half acres of paddy land, as well as some cash money and food grains for 6 months.[9] In 1982, under General Ershad's regime, a third phase was authorised under which 250,000 Bangalis were to be settled in the CHT.

After 1983 the settlement programmes still continued. Officially the low population density in the hills and the overpopulation in the plains were given as an argument for the settlement of Bangalis, invalidly comparing the hilly land in the CHT to the fertile plains. But the underlying motive of these transmigration programmes was to outnumber the *Jumma* people and Bangalise the area to undermine the resistance movement and their demand for an autonomous region. Thus, poor landless Bangalis were used as a weapon to outnumber the *Jumma* population. Apart from the settlement of Bangalis many more Bangalis illegally occupied *Jumma* people's land and often even acquired false ownership documents for it.

As a reaction to the militarization and Bangalisation of the CHT the *Shanti Bahini* also started carrying out attacks on Bangali settlers, trying to drive them out and to prevent more settlers from coming to the CHT. In the name of combating the *Shanti Bahini* and flushing out the insurgents, the army carried out large-scale operations against the civilian population. *Jumma* people were driven out of their homes and their villages and their houses were set on fire. Many of them fled to India or into the jungle. Bangali settlers were then settled on the land that the *Jumma* people had left behind. This was a planned strategy: to drive out the Jumma population from their land and settle Bangalis from the plains on the land the *Jummas* had left behind. Simultaneously this would cut off the *Shanti Bahini* from its support base.

Bangalis who could not be settled on *Jumma* people's land were settled in 'cluster villages', usually next to a military camp, where they served as a protective shield for the military against the *Shanti Bahini*. For their survival they depended on government for food rations, distributed by the army. Till today many Bangalis live in cluster villages and remain dependent on food rations. Despite statements from the government after the Peace Accord that food rations to Bangalis in cluster villages would be stopped rations still continue. Settlers in the cluster villages live in miserable conditions; they are vulnerable and most susceptible to manipulations of the army and better-off Bangali settlers. They have nowhere else to go as they have no land or other means of living anywhere. Settlers from cluster villages have repeatedly told the CHT Commission that they were willing to go anywhere outside the CHT if they were given any means of living that would enable them to survive. However, so far the government has refused to start planning for the resettlement of Bangalis in the plains, even though the EU Parliament had adopted an amendment to earmark part of the aid to Bangladesh "for the repatriation of Bangali settlers in the CHT back to the plains" as early as 1996 (Arens & Chakma 2010:25).

Although the government settlement programmes officially stopped in 1985 Bangalis are still being settled in the CHT and the influx of Bangalis continues till today. The composition of the population has changed dramatically since the 1970s. According to official census figures, the percentage of Bangalis in the CHT rose from 19 percent in 1974 to 38 percent in 1981 and to 49 percent in 1991 (Arens 1997:53). By now the Bangalis have outnumbered the indigenous population.

The situation has become even more complicated in the mean time as many settlers have lived in the CHT for several decades now, many of them have been born there and Bangali businessmen have accumulated large economic interests and become an organised interest group that strongly opposes the peace accord. The main Bangali settler organisations are *Somo Odhikar Andolon* (Equal Rights Movement) and its student wing *Parbatya Chattogram Chattro Parishad* (CHT Student Council). The *Somo Odhikar Andolon* was set up in 2003 with the strong support of Abdul Wadud Bhuiyan, MP of BNP and chairman of the CHT Development Board (CHTDB) at that time. Its leaders are mostly better-off Bangalis who have gained financial interests in the CHT through lease or (illegal) purchase of land or their businesses. *Somo Odhikar Andolon* is patronized by the army and is known for stirring up communal tensions in the CHT. They also have close ties with the *Jamaat-e-Islami* that propagates political Islam and wants introduction of the *Sharia* law in Bangladesh. Besides, Bangalis have become a major voting bank for national political parties. *Somo Odhikar Andolon* reject the Peace Accord because it gives preference to the indigenous people. They demand equal rights for Bangalis, totally negating the historical injustices that the indigenous peoples suffered. *Somo Odhikar Andolon* patronises and mobilises poor settlers who live in cluster villages for their own political interests and the latter have started using the same language, such as "We want equal rights for tribal and non-tribal in CHT" and "No withdrawal of army from the CHT". Whereas before many of the settlers who emphasised that they were willing to go anywhere in the country if they were given means of existence, now tell that they want equal rights, that they want the army to stay in the CHT and that the CHT is their home. These poor settlers are not only being used against the *Jumma* people by the government and the army, but also by *Somo Odhikar Andolon* and other Bangali organisations for their own vested interests.

Land is the biggest source of conflict in the CHT. Bangalis have been settled on land belonging to *Jumma* people and many of them were even given official documents of ownership, adding to the conflict as in most cases *Jumma* owners hold documents for the same plots of land. Settlers have also illegally occupied land on their own. When the *Jumma* refugees returned from India after the Peace Accord they found their land occupied by Bangalis. In many cases this has led to violent conflicts. According to the Peace Accord the returnee refugees are to be rehabilitated on their own land, but in 2010 in total 9,780 out of 12,222 repatriated *Jumma* refugee families still did not get their land back and 40 villages of the returnees are still under occupation of settlers. Besides, more than 90,000 internally displaced families have not yet been rehabilitated (Kapaeeng Foundation 2010:95). The Peace Accord also provides for a Land Commission to resolve all the land conflicts, but so far it has not functioned properly. The government has delayed amendments to the CHT Land Dispute Resolution Commission Act of 2001 as per the proposals of the CHT Regional Council to bring the Act in line with the Peace Accord. Besides, the indigenous members have boycotted the Land Commission because the present (November 2012) chairman of the Land Commission, Former Justice Khademul Islam, turned out to be not impartial and controversial. He unilaterally announced to carry out a cadastral survey in contradiction to the Peace Accord, which specifies a survey to be held after the land disputes had been settled.

Land grabbing by Bangalis is continuing. In 2011 the settlers grabbed at least 7118 acres of land of indigenous owners (IWGIA 2012:341). Land has become scarce and many *Jumma* people have lost their livelihood.

Apart from Bangali settlers who have occupied indigenous people's land, the security forces have also acquired vast areas of *Jumma* people's land, even without giving proper compensation, particularly in Bandarban. The largest acquisitions are approximately 30,000 for training facilities of the army, 26,000 acres for training facilities of the air force and approximately 9,560 acres for expanding the Ruma cantonment (Adnan 2011:58).

'Development'

Foreign interference, in particular development aid, has both directly and indirectly not only added to continuing militarization of the CHT and with that massive human rights violations, but also to a systematic destruction of the mode of production, environment, way of life, culture of the *Jumma* peoples.

From the 1950's the Pakistani government had started opening up the hills. With financial aid from foreign donors development projects were set up in the CHT, such as the Karnaphuli Paper Mill (requiring millions of tons of bamboo and soft wood), the Karnaphuli Rayon Mill and the Kaptai Hydro-electric Dam. The lake formed by the Kaptai dam inundated 40 percent of the arable land in the CHT and displaced more than 100,000 people, about one-fifth of the indigenous population at that time. Most of the displaced people did not get proper compensation and 40,000 of them left to neighbouring India while the rest mostly had to survive by *jum* cultivation. This seriously increased the pressure on the land. The economic exploitation of the CHT and destruction of the *Jumma* peoples' livelihoods started with these developments, only to be continued with even greater force from 1976 when General Ziaur Rahman came to power through a military coup.

General Zia declared that the problems in the CHT stemmed from the underdevelopment of the "backward tribal" area. With financial assistance from the Asian Development Bank, the Chittagong Hill Tracts Development Board (CHTDB) was set up to carry out large-scale 'development' programmes. Main elements of these programmes were road construction, telecommunication and 'upliftment' of the indigenous people who, according to the Khagrachhari Brigade Commander in 1990[10] were "still living in the stone-age". In reality these 'development' programmes were clearly instruments of counter-insurgency. Roads and telecommunication were both essential for counter-insurgency operations and by resettling the indigenous people they could be kept under the control of the army. To eradicate the "primitive and environmentally damaging" *jum* cultivation the *Jumma* population was to be settled in so-called 'model' villages. The argument that swidden (*jum*) cultivation is detrimental to the environment has been used by many governments to cover up their real intentions. That the hidden aim of these 'development' programmes was counter-insurgency became even more evident when the General Officer Commanding (GOC) of Chittagong Division was made ex-officio chairman of the CHTDB in 1982 during the regime of General Ershad (1982-1990).

The Upland Settlement Scheme, which started in 1979, was part of the multi-million dollar Multi-Sectoral programme funded mainly by the Asian Development Bank and UNDP. Under this scheme thousands of reluctant *Jummas* were driven out from their villages in Khagrachhari District through military operations and forcefully relocated in cluster villages under control of the military (similar to the 'strategic hamlets' during the war in Vietnam). Their livelihood and villages were destroyed and they were made to work as day labourers on rubber plantations with the promise that each family would be given four acres of land for rubber plantation, 2 acres for horticulture and 0.25 acres for homestead. According to the Agartala-based Humanity Protection Forum there were more than 200 cluster villages in Khagrachhari District by 1990, including the cluster villages for Bangalis and by 1992 in total an estimated 300,000 *Jumma* people and 200,000 Bangalis had been shifted to cluster villages.[11] This was more than half of the population in the CHT at that time (Arens 1997:63). An army officer frankly told the CHT Commission in 1991:

> The main aim of the cluster villages is to cut the line of supplies to the *Shanti Bahini* and to bring the tribals and Bangalis into the modern line" (CHT Commission 1991:45).

What the latter meant in practice was the transformation of a partially self-sufficient indigenous economy into a dependant market-economy in which the *Jumma* people were made into predominantly cheap wage labourers on plantations and in so-called 'afforestation' programmes where they were more easy to control.

These so-called 'development' programmes, combined with the massive militarization and Bangalisation of the CHT resulted in an escalation of the conflict between the Bangladesh government and the *Jumma* people and the perpetration of massive human rights violations. To enforce the relocation of the *Jumma* population in cluster villages *Jumma* villages were attacked in large-scale military operations under the guise of operations against the *Shanti Bahini*. People were arrested, tortured, killed, and their houses were burnt. Particularly in the 1980s there were large-scale massacres and other serious human rights violations, most of them in Khagrachhari District. As a result more than 70,000 *Jumma* people fled to India. Many of those who could not flee to India fled into the jungle. More than 100,000 *Jummas* were internally displaced within the CHT. Bangalis were settled on the land that the *Jummas* left behind, adding to the communal tensions between *Jummas* and Bangalis.

Reports of gross violations of human rights by the military—mass killings, arrests, torture, rape, disappearances, arson, forced relocation, forced marriages to Bangalis, destruction of temples, houses, and whole villages—started coming out of the area from the early 1980's. Under General Ershad some of the worst massacres took place, such as in Khragrachhari District in 1984 and 1986 (see the following box for major massacres).

Major Killings and Atrocities in the CHT

14 December 1971: Members of the Mukti Bahini (freedom fighters) kill 22 Chakmas at Kakicherra, claims German anthropologist Wolfgang Mey (Ghosh 1986). The killings were caused largely for the Chakma Chief Raja Tridiv Roy's support to the Pakistani regime and allegedly harboring Pakistani supporters, Mizos, and later defeated Pakistani military and *Rajakars* (Mohsin 1997).

Early 1977: Four villages—Matiranga, Guimara, Manikchhari and Lakshmichhari—come under attack of the military; 50 people are shot dead and 23 'tribal' women are tortured to death. Most of the homes in villages under attack are burned (Anti-Slavery Society 1984: 59).

24 December 1977: In an army led raid in villages north of Rangamati, many are arrested and detained without trial or legal representation (Anti Slavery Society 1984: 59).

30 December 1977: The village Kukichara is attacked and 12 members of Nathu Chandra Chakma's family are killed and their houses burned down. Homes of Shanti Lal Chakma and Sukra Moni Chakma are also burned down and both of them are severely injured (Anti Slavery Society 1984: 59).

December 1978: About 50 villages are raided in the north of Ruma army camp; most of the houses in 22 villages of Dumdumya *mouza* (no. 150) are destroyed (Anti Slavery Society 1984: 59).

9 January 1979: The villages in the Subalong valley are attacked; a 70-year old woman is burned alive at her house (Anti Slavery Society 1984: 59).

5 March 1979: Two students, Samiran Talukder and Alomoy Talukder and a farmer Rallawa Chakma, all aged 16, were arrested in the village of Gargajyachari and hacked to pieces by troops (Anti Slavery Society 1984: 59).

2 April 1979: In a raid led by Captain Abul Kamal Mahmud in Kunungopara village 25 hill people are killed and houses are burned. Sindhu Kumar Chakma, Arun Kanti Chakma, Anabil Chakma and a few others of the same family are shot dead and burned in the presence of the surviving family members (Anti Slavery Society 1984: 59).

9 April 1979: In Rangmati, 70 hill people including Kalpa Ranjan Chakma, a member of the then ruling Bangladesh Nationalist Party (BNP) are arrested, detained and beaten by the army (Anti Slavery Society 1984: 59).

27 December 1979: Ven. Ajara Bhikkhu (Buddhist monk) and Ven. Bannitananda Bhikkhu of the Buddhist temple at Thakujyama Kalak in the Kachalong valley are reportedly hacked to death by soldiers. Subsequently, Aggavansa Mahathera (Chakma Rajguru) in his statement to the Commission of Human Rights in United Nation Economic and Social Council alleged that the Bangladeshi regime particularly targeted the hill people on the basis of their religious beliefs (The Charge of Genocide 1986: 59).

25 March 1980: Some 200 to 300 hill people are killed and 600 houses burned in Kaokhali Bazar in Kalampati Union. This act of violence with a record of the largest number of killings in the CHT takes place ten days after the Shanti Bahini attacked and killed 22 soldiers of a company under Captain Abul Kamal Mahmud. Army unit commander of Kaokhali Bazar, Captain Kamal Mahmud called the 'tribal' people to gather to attend a meeting on the restructuring of a temple when the army open fire at the gathering.

A riot by the Bengali settlers reportedly followed and 24 villages were attacked. The priests and temples were particularly victimized--nine temples attacked, five of them completely destroyed.

1980: Shanti Bahini claims that 12,000 to 15,000 people are detained illegally by the army during the year (Anti Slavery Society 1984: 60).

26, 27 & 28 June 1981: A government backed riot reportedly takes place in Matiranga-Tabalchhari Police Station area and some 500 hill people are killed and 100 houses burned. Many are hacked to death or burnt alive as villages are razed to the ground (Anti Slavery Society 1984: 63).

June 1981: Amnesty International reports journalist Sunil Kanti De is arrested and tortured for investigating conditions in the CHT. *New Nation* journalist Saleem Samad is also detained without food and water for several days for similar reasons (Anti Slavery Society 1984: 68).

26 June 1983: The army reportedly starts a bombing operation in the Panchhari Police Station area, resulting in the fleeing of the youths for their lives (Anti Slavery Society 1984: 63).

11 July 1983: In Golakpatimachara village, 12 members of two families are shot dead (Anti Slavery Society 1984: 63).

10 August 1983: 100 houses in the Maramaichyachara village and 120 houses in the Jedarmaichyachara village are burned (Anti Slavery Society 1984: 64).

11 August 1983: In Logang village 150 houses are burned. The same day army accompanied by Bengali settlers, attack Tarabanya village and reportedly kill 50 people. They allegedly hack Surendra Tripura (40) and his wife (67) among others, and lift children by their limbs and smash them to the ground (Anti Slavery Society 1984: 64).

May–June 1984, Bhushanchhara: In retaliation to an attack by Shanti Bahini on settlers' village killing at least 47 Bangalis (according to an Amnesty International report: 100), the army kills at least 63 hill people, and burn 400 houses in 13 villages (The Charge of Genocide 1986: 51). As a result of the rising violence 18,000 hill people flee to Mizoram, India (The Charge of Genocide 1986: 69).

22 July 1985: On Longudu upazila election day, the military chosen candidate Abdul Rashid Sarkar, with the support of the soldiers led by Second Lt. Mohsin of Tintilla camp, attacks the hill people's villages of the upazila. The assailants severely beat Rajani Kanta Chakma, the *pahari* and favorite candidate along with others. Killings, rape, and torture of the villagers take place during the attack. Abdul Rashid Sarker is elected chairman.

July 1985-July 1986: Twenty-five military camps are recorded to be actively operating in the CHT to attack villages in 32 different locations (The Charge of Genocide 1986: 73).

29 April 1986: Armed with automatic weapons, the Shanti Bahini guerillas (approximately 200) kill 20 army personnel in an attack in an army camp at Golakpratima in Khagrachhari upazila.

On the same day, Shanti Bahini guerillas attack adjacent Bangali settlements and four Bangladesh Rifles (BDR) (now Bangladesh Boarder Guard-BGB) outposts near Chantilla and kill 32 civilians and 23 BDR personnel (Ghosh 1986).

29 April to 5 June 1986: The Shanti Bahini reportedly kill at least 50 armed forces personnel and 130 Bangali settlers (Ghosh 1986). These attacks are seen as a protest against the forced relocation of the hill villagers to the "protected" also known as "cluster" villages and to create obstruction to polling in the CHT in the general elections (Amnesty International 1986:15).

Starting on May 1, 1986 the army, BDR, *ansars* (paramilitary) and Bangali inhabitants begin retaliation in the Panchhari-Dighinala-Khagracharia-Matiranga area. They reportedly kill 500 hill people and burn 2000 houses

(Amnesty International 1986: 15).

July 1986: President Ershad, in a speech to the members of the parliament, says that during the last 10 years 1,000 civilians including Bangalis and members of the armed opposition have been killed. He also says that 213 members of the security forces have been killed (Amnesty International 1986:10).

August 1988: Thirty-eight hill people are killed and 250 houses are burned in Bagaichhari.

May 1989: Thirty-six hill people are killed and hundreds of houses are burned in Longadu.

2 February 1992: In Malya two bombs are detonated in a passenger launch carrying a delegation of the hill people to Rangamati and Dhaka protesting the misdemeanor of the army. These explosions and the subsequent attack by the Bangali settlers on the survivors who swim to the shore cause 30 deaths.

10 April 1992: The Bangali settlers, with support of the BDR, the Village Defence Party (VDP) and armed police attack villages in Longodu and kill 138 hill people and burn 550 houses. Probe (English magazine) argues that the number is vastly exaggerated, claiming the actual number to be 12. Apparently 20 persons listed among the 138 reportedly dead by the Pahari Chhatra Parishad (Hill Students' Council) physically showed up before the Enquiry Commission (Probe 1994: 40).

17 November 1993 (during the cease-fire period): Twenty-nine hill people and one Bangali are killed and 150 wounded. Bangali settlers reportedly supported by the military attack a peaceful demonstration of CHT Hill Students' Council in Naniarchar, Rangamati Hill District. The assailants also reportedly burn 25 houses.

Between January 1991 and June 1992: Forty-seven hill women are reported abused by Bangali security personnel (Hill Watch Human Rights Forum 1992: 178)

15 March 1995: Three hundred houses are burnt in Bandarban.

12 June 1996: Kalpana Chakma (23), the organizing secretary of the Hill Women Federation, an organization fighting for the rights of the indigenous women in the CHT is allegedly abducted by the military personnel from her house in the village of New Laillaghona (Baghaichhari Polica Station, Rangamati District). (Moral and Salam 2009: 288).

Prepared by Asif khan with Philip Gain

These reports about massive human rights violations made the Swedish and Australian governments withdraw from respectively road construction and afforestation programmes. Except for a few donors, such as the World Bank and the Asian Development Bank, most others refrained from further financing development programmes in the CHT. However, although for many donors one of the criteria for giving aid was (and is) that a receiving government respects international standards of human rights, they did not refrain from giving aid for government programmes in the rest of Bangladesh. This aid thus indirectly contributed to the defence budget and with that to the gross human rights violations committed by the army in the CHT (Arens, 1997:45-80).

After the signing of the Peace Accord in 1997 most donors remained hesitant to fund development programmes in the CHT, but after the UNDP launched a large-scale programme many of them stepped in as well. National NGO's (Non Governmental Organisations) such as BRAC and Grameen Bank also started programmes in the CHT. They mostly replicate the micro-credit programmes that they carry out in the plains, despite strong criticism of the *Jumma* people that those programmes do not fit the situation in the CHT. Indigenous peoples themselves also set up their own NGO's, but they faced and still face a lot of interference by the government and the army. For instance, NGO's in the CHT need clearance from the DGFI (the army intelligence wing) in order to register with the NGO Affairs Bureau (which is necessary to receive foreign funds) (IWGIA report 14, 2012:20). This gives the DGFI full control over these NGOs. They also have to provide information about their projects to the district administration and the army every year. One of the latest directives is that half of the beneficiaries of their programmes should be Bangalis and they have to report how many of their staff members are Bangali and how many indigenous. The government and army are clearly in tune with the settlers demanding equal rights and ignoring the need to repair historical injustices against the indigenous peoples.

Since 2010 the government has put new restrictions on foreign nationals visiting the CHT. It is probably not coincidental that these restrictions were put in place after Lars-Anders Baer had issued his report about implementation of the Peace Accord, as mentioned earlier. Now foreigners are not only required to inform the District Commissioner about their intended visit, but they also have to submit their exact schedule of when and where they plan to meet whom, thus preventing people to meet and talk in confidentiality. The CHT Commission was compelled to discontinue its programme in the CHT in November 2011 due to severe obstruction and interference by government officials and intelligence agents during meetings with civil society groups in Rangamati and Bandarban. These officials insisted on being present despite requests from the commission to leave because this would violate the principles of confidentiality and trust and affect the testimonies of those present. Some of the officials stated that they were under orders from their superiors to accompany the mission in the CHT at all times and referred to written instructions from concerned ministries. As The CHT Commission refused to compromise its work principles it cancelled the meetings and left the CHT.[12]

Donors continue to finance development activities in the CHT, even though serious human rights violations continue, often carried out by settlers backed by the security forces. Below is a list of major human rights violations after 1997. Apart from these there have been numerous other human rights violations.

Major human rights violations after the 1997 Peace Accord

26 August 2003, Mahalchhari: Nine villages in Mahalchhari in Khagrachhari Hill District are allegedly attacked by Bangali settlers with support of the army personnel. Two hill people are killed (one of them an eight month infant), nine women raped, 379 households burned, and at least 46 hill people injured. A Buddhist temple of Babupara founded in 1962 is completely destroyed (Saha and Khanam 2009: 292).

September-November 2006: Two hundred and seventy five families are forcefully evicted for military artillery firing range on 11,444 acres of land that had been acquired by the army between 1990-93.

23 February 2007: Community leader Ranglai Murong is arrested on false charges, sentenced to 17 years, but finally acquitted and released from jail in January 2009.

20 April 2008, Bagaichhari: Two hundred houses in seven villages are burnt in Bagaichhari of Rangamati. Most of the houses belong to the hill people who allege that the Bengalis backed by the army are behind the incident.

20 February 2010, Bagaichhari: Two hill people are killed and 474 houses of which 397 belong to the hill people are burnt down in violent clashes between the Bangalis and the hill people in 11 villages of Sajek union under Bagaichhari upazila of Rangamati.

April 2011: Hatimura area, Hafchari Union, Ramgarh, Khagrachhari District - looting and arson attack by settlers, backed by the army - 97 houses in seven villages burnt down, two *Jumma* died, one went missing and 20 were injured.

Since 1973 to date: Acquisition of 9,560 acres of land continues for Ruma cantonment, despite petitions from *Jumma* leaders to cancel the acquisition. If completed this will result in the eviction of 5,000 families from 13 *mouzas*.

Source: Local leaders, various reports by chtnews.com, PCJSS, Kapaeeng Foundation, CHT Commission, IWGIA, and Society for Environment and Human Development (SEHD) 2008 and 2010.

Prepared by Asif Khan with Philip Gain

There have been no independent investigations in any of the human rights violations. Of the few government investigations carried out no reports have been made public and none of the perpetrators have been arrested, let alone tried in court. A culture of impunity still prevails.

Killings, rape, torture, arbitrary arrests, land grabbing, and burning of houses continue. This has a huge impact on the daily lives of the indigenous peoples and on their culture and way of life.

Impact on the Lives of the People

The decades long militarization, Bangalisation and gross human rights violations have left inerasable scars on the bodies, hearts, and minds of the *Jumma* people, their culture, their society, and their environment. People of all generations have been severely traumatized by the atrocities that they have experienced or witnessed. Many children have lost their parents and have grown up in orphanages or in their extended family. The long-term effects of these traumatic experiences and the implications for future generations cannot be ignored.

On its first visit in December 1990 an old man in Khagrachhari District told the CHT Commission: "The army is following us everywhere. Life is not ours" (CHT Commission, 1991: 89). Now, 22 years later, the army is still everywhere in the CHT and *Jumma* people still do not have control over their own lives, nor over their own land. Most of the indigenous people still live continuously with physical and mental insecurities. They are continuously watched, harassed and intimidated by army personnel and in some areas they cannot stay in their houses. In areas where the UPDF is strong people are regularly questioned and searched by the army. One woman told how the army harassed and intimidated her while she was grazing her cows:

> The following repeated incidents in Sajek union, Baghaihat, Rangamati District are examples of the continuing attacks on the indigenous people.

In the night of 20 April 2008 a group of Bangali settlers led by Selim Bahari and Gulam Molla, two local leaders of Baghaihat branch of *Somo Odhikar Andolon*, attacked seven villages in Sajek Union, and set houses on fire. The settlers were allegedly supported by personnel from 33 East Bengal Regiment of Baghaihat zone led by Lt. Col. Sajid Imtiaz. In total 76 houses of Jumma families were burnt down.[13] As one indigenous person told an Bangali citizens' committee fact-finding mission:[14]

> Last January, our houses were grabbed by the settlers, under the leadership of Somo Andolon leader Golam Mowla. We don't get a chance to speak when council office meetings are held. When the Raja (Devasish Roy) came to visit, the army camp ordered us not to speak to him. On the night of 20th April, it was very hot, and I was sitting outside the house that I had raised after the January attacks. Suddenly I heard some Bengalis shouting "Narae Takbir Allahu Akbar". I could see fire in the distance. I could hear Paharis shouting "Ujo, Ujo" [advance]. At this time, I saw an army vehicle. By then, both my houses and other surrounding houses had caught fire. On the one hand, our houses were burning, while on the other, the settlers were looting.

A Jumma whose house had been burnt down stated:[15]

> The people who were setting things alight, first took out from our homes the TVs, beds, wardrobes; whatever they found, they looted, and at the end they torched the houses. Those who set the houses alight took everything.

This attack should be seen against the background of a renewed programme of the previous (BNP) government to settle 10,000 Bangali families in Sajek union. In 2005 the army constructed a road through the area and settlers were to be settled along that road.[16] *Jummas* had been protesting against the fresh settlements. After the attacks Bangali settlers and the BDR (now BGB) occupied the *Jumma* people's land, mostly their *jum* land that they owned collectively under customary rights. The settlers wrongfully claim that it is *khas* (fallow government) land. It should be noted here that according to indigenous customary laws land is communally owned and is yearly divided by the village headman among the people for *jum* cultivation. The 1997 Peace Accord recognises the customary land rights of the indigenous people.

Jumma people whose lands had been occupied were prevented by the BDR from going to their land and settlers regularly threatened them. Some of the following statements also clearly show the collaboration between the security forces and the settlers. One woman whose land was occupied told:

> The tensions in Sajek did not decrease after that. On 19 and 20 February 2010 there was another, even larger, attack on *Jumma* villages in the area by settlers supported by army personnel. This happened after *Jummas* had prevented new settlers from building houses on their land. In total 434 *Jumma* houses, two Buddhist temples, one church and two schools were burnt down in 13 villages and two *Jumma* men were shot dead (PCJSS 2010). There have been no official investigations into any of these attacks and no perpetrators have been arrested or tried in court.

Like in Sajek many other indigenous people have no security of life and live in poverty as they have lost their land. For many of them there is no more land to practice *jum* cultivation and they only depend on daily wages, petty business or other means of survival. The case of the artillery firing range in six *mouzas* in Bandarban District illustrates this.

From 1991 the Army acquired 11,444 acres of land for an artillery firing range bit by bit and started evicting people from their homesteads and agricultural land without proper compensation. Mru people mostly inhabited the land and there were also some Chakma and Bangali families who had been resettled there from Kaptai in the 1960s after the construction of the Kaptai Dam. There were no settlers in the area, but Bangali lease takers have come in. As per an agreement between the Union

Council chairman Ranglai Mru and the army some people who owned land under the artillery range were allowed to stay and lease their own land from the army. In 2006 the army suddenly increased the lease fee tenfold. None of the *Jumma* and Bangali cultivators could pay the high lease fee and then the army forcefully relocated 275 families at gunpoint. People then protested against the artillery range; Ranglai Mru, the Union Council Chairman, demanded proper rehabilitation of the people. In February 2007, during the caretaker government, Ranglai Mru was arrested by the Army on false charges of illegal possession of a pistol and convicted to 17 years imprisonment. He was tortured, kept in shackles and not given proper medical treatment after he suffered a heart attack in jail. After strong protests by the National Human Rights Commission, Ain 'O Shalish Kendra, and Amnesty International Ranglai Mru was finally released in January 2009 (IWGIA report 14, 2012).

According to the local people the army officially acquired 11,444 acres, but in fact they took 50,000 to 60,000 acres, all land under customary rights. The evicted people, both indigenous and Bangali, lead a miserable life; they have no more food security. They cannot cultivate their land because of the shooting practices. Several elderly people and children died. There are no medical facilities and many live in makeshift houses on the roadside. They have to live on daily wages, collecting and selling wood. The people demand that the army return the extra land to the people and let them use those 11,444 acres for the 9 months of the year when the army is not using it as a shooting range.

The army leases out a lot of the land to companies for agro-forestry and pisciculture. Some companies that have leased large areas of land are Destiny, Exim Group, Meridian Group, Imam Group, and Mustafa Group. Several people told they didn't even know that their land had been signed away. Bypassing the headman these companies got leases to their land. Sometimes landowners just find strangers measuring their land or planting trees on their land and then it turns out that their land has been signed away. On top of that some of the people whose lands have been taken have been implicated in false cases by the land grabbers. Companies have also illegally bought thousands of acres of land in the CHT, such as the Islamic organisation Muhammadiya Jamia Sharif (known as the Laden group) in Lama *upazila* and Naikhongchhari *upazila* in Bandarban district (Kapaeeng Foundation 2012).

Not only the livelihoods, but also the cultures of the indigenous peoples, their language, collective spirit, dress, festivals, cultural and religious expression, are slowly being destroyed. *Jumma* people expressed that this has resulted in the erosion of many traditional values and that their society has lost its harmony and traditional balance (Mohsin 2012). They are becoming more individualistic, the collective spirit is eroding, and class divisions are increasing (Roy a.o., 2010:226). This is largely due to the situation of domination by the security forces and Bangalis. Many Bangalis look down on the indigenous way of life and cultures as "backward", call the indigenous languages "dialects" and see their cultural expressions only as exotic commercial objects to entertain outsiders or to exhibit as artefacts in museums, rather than as an essential part of the indigenous ways of life. Many young women do not feel safe to wear their traditional dress for fear of rape or harassment. Education is lagging behind, especially for the smaller indigenous groups and the drop-out rate is high. Apart from poverty an important reason for this is that the children are being taught in Bangla and many children do not know Bangla at all when they start primary school. Besides, the curriculum is tuned to the situation and culture of Bangalis in the plains (Roy, a.o., 2010:242).

Women As Specific Targets

Before the Peace Accord the security forces have particularly targeted women and rape has been used systematically as a weapon of counter insurgency. Rape was a recurring characteristic of attacks on *Jumma* villages and women were often [gang]raped in front of their husbands, children and other family members. Young women were kept by the army in their camps for days and raped. The indigenous Hill Women Federation stated in 1995 that over 94% of the rape of *Jumma* women in the CHT between 1991 and 1993 were by security forces and over 40% of the victims were women under 18 years of age (CHT Commission 1997:12). Women who protested against human rights violations were not safe. The case of Kalpana Chakma, 23-year old organising secretary of the Hill Women's Federation who had repeatedly protested against attacks on *Jumma* villages in her area is widely known. She was abducted from her home on the night of 11 to 12 June 1996 allegedly by army personnel led by Lieutenant Ferdous, the commander of the army camp near her village in Khagrachhari District. No sign of Kalapana Chakma has been seen or heard since. There has never been an official independent investigation and no one has ever been arrested.

Women are seen as the procreators of the group and targeting women shows the deliberate intent to extinguish the whole ethnic group and its future generations. This is also what happened during the Bangladesh liberation war when the Pakistani army raped over 200,000 Bangali women. The exclamation of a West Pakistani soldier: "We are going. But we are leaving our seed behind", as quoted in Sharlach (2000:95), is strikingly similar to that of the Bangali soldiers who exclaimed: "No Chakmas will be born in Bangladesh" during the Barkal massacre in 1984 in which some 200 people were killed and many women raped (Amnesty International 1986:14).

Rape has not only been used during counter-insurgency. Also after the Peace Accord rape occurs frequently, mostly by Bangali settlers, but also by security forces. Women still live in continuous fear of rape when they go to their *jum* field or the market, go to collect firewood, when they have to pass by a Bangali village or an army camp; and even at home they are not safe. *Jumma* Net (2009), a Japan based organisation, reported 15 rape cases with 26 victims and five attempted rapes with five victims between 2003 and 2006; in four of the 15 cases the rapists were Army or BDR personnel. There have been reports of attempted rape of women by army personnel as well. Women who work in offices with Bangali colleagues are also not safe. A report by Raja Devashish Roy a.o. (2010:240) mentioned that a Marma woman was raped by her Bangali colleague at her office. The same report mentioned more rape cases. Settlers, backed by the army during the Mahalchhari incident in 2003, raped eight indigenous women; settlers raped two minor girls during the Maischhari incident in 2006, and in 2009 another Bangali raped a three-year old Chakma girl. In 2011, settlers reportedly raped 11 indigenous women from the CHT and five of them were killed after rape. Besides there were six rape attempts, one of these by security forces personnel and five women were abducted by settlers (IWGIA 2012:342). In 2012 about eight *Jumma* women were reportedly raped by settlers, one of them was killed after rape and there had been two attempted rape cases.[17] (Also one Bangali woman in the CHT was reportedly raped by a settler and two rape cases of indigenous women in the north of Bangladesh were reported) Several of the rape victims were mere children, including a 3-year old girl. The girls and women who survived are traumatised for the rest of their lives. Three recently reported cases.

a. On 28 April 2012 Ms. Alpana Chakma alias Kalabi (45), wife of Kaladhan Chakma of Jarulchara, Gulshakhali union, Longadu upazila, Rangamati district was allegedly assaulted and raped by Bangali settlers on her way home from Mieni Doar Bazaar. After that the rapists attempted to kill her and left her unconscious, believing she was dead.

b. On 9 May 2012 Sujata Chakma (11), daughter of late Mr. Jyotish Chandra Chakma and Ms. Mongala Devi Chakma of Ultachhari *mouza*, Atarokchhara union, Longadu upazila, Rangamati district was killed after rape by a Bangali settler when she was grazing cows half a kilometer from her village. A six-year old child who was with her reported the news. Longudu police arrested the alleged rapist and killer from his house at Uttar Yarengchhari, Atharokchhara union the next day. So far there has been no report of a conviction of the rapist.

c. On 5 June 2012 a Tripura girl (12) was raped by two settlers at Baladhuram Para, Taindong union, Fatikchari upazila, Chittagong district. When she was alone at home truck driver Md. Ramzan (35) and helper Md. Hasan raped her. Villagers rescued the girl when they heard her cry.

Many of the rape cases go unreported for fear of women to be ostracised or rejected by their husbands, but several indigenous womenís organisations have started speaking out against rape and supporting women who have been raped. Women are encouraged to report rape and they are supported in the medical and legal processes. So far hardly any of the rapists have been arrested, let alone sentenced. This stands in sharp contrast to some severe jail sentences and even death sentences of rapists in the plains.

Women have been lobbying since long to get rape acknowledged as a war crime. Finally, on 19 June 2008, the United Nations Security Council adopted Resolution 1820 and noted:

"Rape and other forms of sexual violence can constitute a war crime, a crime against humanity, or a constitutive act with respect to genocide"

The Resolution urged the UN to impose sanctions on violators. It is high time that the government takes serious steps to severely punish rapists in the CHT, both civilian and [para]military.

Conclusion

In order to resolve the CHT conflict it is first and foremost essential that the government and Bangali society acknowledge the historical injustices committed by subsequent governments against the indigenous peoples. To make such an acknowledgement concrete the most urgent tasks would be constitutional recognition of the identity of the indigenous peoples and their rights, withdrawal of the army from all the still existing temporary camps to the six cantonments in the CHT as per the Peace Accord, amendment of the CHT Land Dispute Resolution Commission Act of 2001 to bring the Act in line with the Peace Accord, appointment of an impartial chair of the land commission, resolution of all land dispute and a halt to a further influx of Bangalis. Besides, as long as the culture of impunity still prevails, human rights violations will continue to go unpunished. Thorough investigation of human rights violations in the past and in the future by independent bodies are a must, reports of the investigation should be published and the perpetrators should be identified, arrested and tried in court.

Notes

[1] Aditya Kumar Dewan: 'Class and Ethnicity in the Chittagong Hill Tracts of Bangladesh', 1991, unpublished PhD thesis, chapter IV: Colonial transformation of indigenous systems.

[2] idem.

[3] The international CHT Commission was set up in 1989 and re-established in 2008 with a renewed mandate to promote respect for human rights, democracy and restoration of civil and judicial rights in the CHT, including examination of the implementation of the CHT Accord of 1997. The re-established CHT Commission consists of international and Bangali members.

[4] The stated number of camps that have been withdrawn after the 1997 Peace Accord varies heavily depending on the source. According to PM Sheikh Hasina in November 2011, 235 camps have been withdrawn (http://www.thedailystar.net/newDesign/news-details.php?nid=211364), whereas Shantu Larma, PCJSS leader and Chairman of the Regional Council, claimed that only 74 camps and one army brigade that was stationed in Rangamati cantonment had been withdrawn (IWGIA report 14, 2012:12).

[5] From an interview by Jérémie Codron with an army officer in 2003.

[6] Betsy Hartmann and Hilary Standing: 'The Poverty of Population Control. Family Planning and Health Policy in Bangladesh', Bangladesh International Action Group, London, 1989, p. 10.

[7] *New Age* June 8, 2012

[8] Various reports in Daily Star in April/May 2012. See e.g. http://www. thedailystar.net/newDesign/news-details.php?nid=232544

[9] Anti Slavery Society: 'The Chittagong Hill Tracts. Militarization, Oppression and the Hill Tribes, London, 1984, Appendix 1.

[10] Personal communication of Khagrachhari Brigade Commander Sharif Aziz to the CHT Commission on its first visit to the CHT in 1990 (CHT Commission 1991:65).

[11] Humanity Protection Forum: 'On Memorandum to Governments and Institutions Giving Aid to Bangladesh, February 1992, p. 9.

[12] Press Release: CHT Commission concludes Sixth Mission 30 November, 2011. See: chtcommission.org.

[13] See Kapaeeng Watch report at: http://groups.yahoo.com/group/muktomona/message/48000 (accessed 24 June 2012).

[14] Fact Finding Team 1 (Moshrefa Mishu, et al) Report on 20th April 2008 Incident at Sajek Union. At: http://www.drishtipat.org/blog/wp-content/uploads/2008/05/sajek2.pdf (accessed 17 July 2012).

[15] Press statement Sajek's burnt villages: Citizen's team calls for inquiry and urgent relief. 5 May 2008. See: http://mail.sarai.net/pipermail/reader-list/2008-May/012803.html (accessed 17 July 2012).

[16] See JSS report at: http://pcjss-cht.org/Major%20Communal%20Attack/10-Baghaihat-Khagrachari%20Attack%20_19-23%20Feb%202010_.pdf (accessed 24 June 2012).

[17] Reported by Kapaeeng Foundation and chtnews.com.

References

Adnan, Shapan and Ranajit Dastidar. 2011. *Alienation of the Lands of Indigenous Peoples in the Chittagong Hill Tracts of Bangladesh.* Dhaka: CHT Commission, Copenhagen: IWGIA.

Amnesty International. 1986. Unlawful Killings and Torture in the Chittagong Hill Tracts. London: Amnesty International Publications. 38 pp.

Arens, Jenneke. 2011. Indigenous Peoples and Genocide: The Chittagong Hill Tracts, Bangladesh. In: Samuel Totten and Robert K. Hitchcock (eds) *Genocide of Indigenous Peoples. Genocide: A critical Bibliographic Review, Volume 8.* New Brunswick and London: Transaction Publishers.

Arens, Jenneke and Kirti Nishan Chakma. 2010. "Indigenous Struggle in the CHT." In: Mohaiemen, Naeem *Between Ashes and Hope: Chittagong Hill Tracts in the Blind Spot of Bangladesh Nationalism.* Dhaka: Drishtipat Writers' Collective.

Arens, Jenneke and Kirti Nishan Chakma. 2002. Bangladesh: Indigenous Struggle in the Chittagong Hill Tracts. In Mekenkamp, Monique, P.Van Tongeren, H. Van de Veen (2002) *Searching for Peace in Central and South Asia.* Boulder, London: Lynne Riener Publishers.

Arens, Jenneke. 1997. Winning Hearts and Minds: Foreign Aid and Militarisation in the Chittagong Hill Tracts. In Economic and Political Weekly Vol.32, nr.29, Mumbai, India.

Arens, Jenneke. 1997. Foreign Aid and Militarisation in the Chittagong Hill Tracts. In Bhaumik, Subir, Meghna Guhathakurta, Sabyasachi Chaudhury *Living on the Edge: Essays on the Chittagong Hill Tracts.* Katmandu: South Asia Forum For Human Rights.

Bhaumik, Subir, Meghna Guhathakurta, Sabyasachi Chaudhury. 1997. *Living on the Edge: Essays on the Chittagong Hill Tracts.* Katmandu: South Asia Forum For Human Rights.

Chakma, Mangal Kumar (ed.). 2011. *UNPFII Study on the Status of the CHT Accord of 1997 and Statements delivered at the UNPFII's 10th session on the said Study.* Dhaka: Kapaeeng Foundation, ALRD.

Chittagong Hill Tracts Commission. 1991. *Life Is Not Ours: Land and Human Rights in the Chittagong Hill Tracts, Bangladesh.* Copenhagen: IWGIA and Amsterdam: Organising Committee Chittagong Hill Tracts Campaign

Chittagong Hill Tracts Commission. 1992, 1994, 1997, 2001. *Life Is Not Ours: Land and Human Rights in the Chittagong Hill Tracts, Bangladesh,* Update 1, 2, 3, 4. Copenhagen: IWGIA and Amsterdam: Organising Committee Chittagong Hill Tracts Campaign

Codron, Jérémie. 2007. Putting Factions 'Back in' the Civil-Military Relations Equation: Genesis, Maturation and Distortion of the Bangladeshi Army', *South Asia Multidisciplinary Academic Journal.* URL:URL: http://samaj.revues.org/document230.html (17 May 2012).

Dewan, Aditya Kumar. 1991. *Class and Ethnicity in the Chittagong Hill Tracts of Bangladesh.* Unpublished PhD thesis.

Guhathakurta, Meghna. 2004. Women negotiating change: The structure and transformation of gendered violence in Bangladesh. Cultural Dynamics 16(2/3), 193-211.

IWGIA. 2012. *The Indigenous World.* Copenhagen: International Work Group for Indigenous Affairs (IWGIA).

IWGIA. 2012. *Militarization in the Chittagong Hill Tracts Bangladesh: The slow demise of the region's indigenous peoples.* IWGIA report 14. Amsterdam: Organising Committee CHT Campaign, Tokyo: Shimin Gaikou Centre and Copenhagen: IWGIA.

Kapaeeng Foundation (a). 2012. Huge land grabbing by outsiders in Bandarban evicting indigenous villagers and permanent Bengali residents. See: See: http://kapaeeng.org/huge-land-grabbing-by-outsiders-in-bandarban-evicting-indigenous-villagers-and-permanent-bengali-residents-3/.

Kapaeeng Foundation (b). 2012. Human Rights Report 2011 on Indigenous Peoples in Bangladesh. Dhaka: Kapaeeng Foundation.

Kapaeeng Foundation. 2010. Human Rights Report 2009-2010 on Indigenous Peoples in Bangladesh. Dhaka: Kapaeeng Foundation.

Levene, Mark. 1999. The Chittagong Hill Tracts: A case study in the political economy of 'creeping' genocide. Third World Quartely, 20(2), 339-369.

Mey, Wolfgang, ed. 1984. *They Are Now Burning Village After Village. Genocide in the Chittagong Hill Tracts.* IWGIA Document 51, Copenhagen: International Work Group for Indigenous Affairs (IWGIA).

Mohaiemen, Naeem. 2010. *Between Ashes and Hope: Chittagong Hill Tracts in the Blind Spot of Bangladesh Nationalism.* Dhaka: Drishtipat Writers' Collective.

Mohsin, Amena. 2012. Bangladesh: Patriarchal State And *Jumma* Women's Agency. See: http://www.sacw.net/article2574.html.

PCJSS. 2010. Report on Massive communal attack on *Jumma* villages by military forces and Bengali settlers in Baghaihat and Khagrachari. http://pcjss-cht.org/Major%20Communal%20Attack/10-Baghaihat-Khagrachari%20Attack%20_19-23%20Feb%202010_.pdf

Roy, Raja Devashish. 1994. Patterns of Land Use in the Chittagong Hill

Tracts. Prospects and Problems, Dhaka.

Roy, Raja Devasish, Pratikar Chakma, Mong Shanoo Chowdhury, Mamong Thuai Raidang. 2010. Hope and Despair: Indigenous *Jumma* Peoples Speak on the Chittagong Hill Tracts Peace Accord. Philippines: Tebtebba.

Sharlach, Lisa. 2000. Rape as Genocide: Bangladesh, the Former Yugoslavia, and Rwanda. New Political Science, 22(1), 89-102.

Documents consulted on Major Killings and Atrocities in the CHT and Major human rights violations after the 1997 Peace Accord

Amnesty International. "Unlawful Killings and Torture in the Chittagong Hill Tracts". London: Amnesty International Publications. 1986. Print.

Mohsin, Amena. "Politics of Security II: Bangladesh Period". The Politics of Nationalism. Dhaka: University Press Limited, 1997. Print.

Kamaluddin, S. A Tangled Web of Insurgency. Far Eastern Economic Review. 23 May. 1980 in Mohsin, Amena. "Politics of Security II: Bangladesh Period". The Politics of Nationalism. Dhaka: University Press Limited, 1997. Print.

Foreign Aid and Militarization in the Chittagong Hill Tracts by Jenneke Arens. Living on the Edge: Essays on The Chittagong Hill Tracts. Edited by Subir Bhaumik, Meghna Guhathakurta, and Sabyasaschi Basu Chaudhury. South Asia Forum for Human Rights. 1997.

Restricted. n.d. "Parbattya Chattagramer Birajman Poristhiti".unpublished paper. Dhaka: AHQ. in Mohsin, Amena. "Politics of Security II: Bangladesh Period". The Politics of Nationalism. Dhaka: University Press Limited, 1997. Print.

The CHT Commission. "Life is not Ours". Land and Human Rights in the Chittagong Hill Tracts Bangladesh. Denmark: Netherlands, 1991.

The CHT Commission. "Life is not Ours". Land and Human Rights in the Chittagong Hill Tracts Bangladesh. Denmark: Netherlands, 1992 in Mohsin, Amena. "Politics of Security II: Bangladesh Period". The Politics of Nationalism. Dhaka: University Press Limited, 1997. Print.

Hill Watch Human Rights Forum. Chittagong Hill Tracts "A Land of Blood and Tears": An Account of Human Rights Violations in the Chittagong Hill Tracts, Bangladesh. 1992. in Mohsin, Amena. "Politics of Security II: Bangladesh Period". The Politics of Nationalism. Dhaka: University Press Limited, 1997. Print.

Anti-Slavery Society. 'The Chittagong Hill Tracts: Militarization, oppression, and the hill tribes. Series. 2. London: Indigenous People and Development. 1984.

The Charge of Genocide: Human Rights in the Chittagong Hill Tracts of Bangladesh, "Papers for the Conference on the Chittagong Hill Tracts". Amsterdam. 1986

Moral, Shishir, and Salam, FMA, "Kalpana Chakma Remains Untraced", Investigative Reports: Environment and Human Rights. Ed. Philip Gain. Dhaka: SEHD. 2009. Print.

Saha, Partha Shankar and Khanam, Khadiza with Gain, Philip. "Burning, Killing and Rape in the CHT: Indigenous Peoples in Despair". Investigative Reports: Environment and Human Rights. Ed. Philip Gain. Dhaka: SEHD. 2009. Print.

Saha, Partha Shankar. "Who Set Fire in the Hills? "Only God Knows"". Investigative Reports: Environment and Human Rights. Ed. Philip Gain. Dhaka: SEHD. 2009. Print.

Probe "Life in the Chittagong Hill Tracts. Dhaka: BCDJC. 1994.

Ghosh, Santanu. Civil War in Chittagong. Sunday (India) 6-12 July 1986: 32-35 in The Charge of Genocide: Human Rights in the Chittagong Hill Tracts of Bangladesh. Amsterdam October 11, 1986.

Additionally local leaders, various reports by chtnews.com, PCJSS, Kapaeeng Foundation Kapaeeng Foundation, CHT Commission, IWGIA were consulted.

Antahapara Army Camp: Impacts on Life and Nature

Partha Shankar Saha

Boom, boom, boom! The silence in the winter morning breaks with the sounds of firing. A stranger to Antahapara, a small hamlet in the Chittagong Hill Tracts (CHT), I shiver in fear. As I walk on the hilly path, I suddenly stop. However, my local guide Anil Kumpai Tripura (22) remains passive.

"This is army firing; I am used to it from my childhood," says Anil.

Red flags around the army camp send alert signals of the firing practices.

The Antahapara army camp in Roangchhari upazila is one of many in Bandarban Hill District. Around 140 families—84 Tripura, and 56 Tanchangya—live at Anthahapara, 20 km north-east of the district town.

Octogenarian *karbari* of the hamlet, Koila Chandra Tripura says the army camp was set up in 1981 during the insurgency of the Shanti Bahini (the armed wing of the PCJSS).

The Shanti Bahini was active in Anatahpara at that time enticing the military government to set up the army camp to 'protect the hill people' in Noapatang union. The camp is located at a strategic point enabling the army personnel to keep vigil on all three villages— Antahapara in the west, Gokhongpara in the east and Kanaijupara in the north. The army camp under the Bandarban Sadar Zone is tasked to watch 20 villages nearby. Bagmara Army Camp is another temporary camp in Noapatang.

The land (10 acres) taken for the Antahapara Army Camp belongs the family of Shambhunath Tanchangnya, the Noapatang chairman. "It is our leased land. The army forcibly set up the camp. We had a big teak garden on this land; the army cleared all trees," informs Shambhunath.

Shambhunath's family used to get Tk.500 annually year from

the camp. Later on the amount was increased to Tk.750. The family stopped collecting this meager annual cash pay from the camp since 2009.

There is a vast plantation of pulpwood under the Kaptai Pulpwood Division under the Forest Department (FD) adjacent to Antahapara. Kaptai Pulpwood Division is one of two divisions that supplies pulpwood to the Karnaphuli Paper Mill. Dhonbadu Tripura and Kermoti Tripura of Antahapara complained that they lost their land to the pulpwood plantation. They applied to the FD to get their land back with no result. Instead, the FD officials threatened to hand them over to the military.

The residents of Antahapara inform that the army personnel assist the FD and the KPM authorities to harvest wood from the forest and pulpwood plantation.

The aged people in the village inform that the army used the villagers to clear the forest and set-up the camp. Pay for their labor was little.

"The army made us slaves. Whenever they wished they took us to the camp and made us work," complained an elderly villager who agreed to talk on condition of anonymity. The villagers inform that they were made to provide with anything the army wanted— bamboos, trees, cows, hens, etc. The army, however, paid for those. "Of course they would decide the prices. The villagers never dared protesting," says Jeromeo Tripura (38).

Almost all of the adult men of Antahapara this writer talked to had experience of being tortured by the army. Shombhunath Tanchangya, a suspect of being associated with Pahari Chhatra Parishad (a student wing of PCJSS) was physically assaulted by the army.

"The young students were the army's main target. The army suspected any student of engaged with political activities," informs Shmbhunath.

"The army first beat me up when I was in class six. These are painful memories I can't forget," says Jeromeo.

"The surroundings started to change right after the army camp was set-up in my village. We used to walk around freely before. But right after the army arrived here, even in my childhood, my movement was obstructed. The difficulties grew as I grew up," says Sreemoti Tripura (39), primary school teacher in the village. "We had to tolerate the rough words from the army off and on. Sometimes I protested their misbehavior," says Sreemoti.

"Before the camp was set-up, I used to go to the jungle by myself. But now girls do not move alone; they go in groups," says Kermoti Tripura (48), another woman of the village.

The villagers report that in 2000 when a Bengali was abducted, all the villagers including the children, were locked up in the army camp.

The villagers in Antahapara inform that the harassment was at its peak during the insurgency. For example, if a visitor came, the villagers had to report it to the camp; they were called to the camp if their shopping was little more than usual; if any dog barked at night, the army entered the village; if anyone kept the light on late at night, it was an offense; if anyone used a torch light at night, one had to explain it to the camp; the list can be longer.

"The army camp was set-up for their own security, not that of the people. The people suffered after the camp was set up," says Shombhunath.

However, the villagers inform that the harassment decreased after the Peace Accord was signed in 1997. Now nobody is called to give free labor in the camp. The army does not frequently enter the villages. We can walk freely now compared to before. There is a volleyball court at the foot of the camp hill. The army personnel play with the villagers on that court.

"But the fear of the army is still there in our blood. If an army person enters the village, a child still screams 'army came, army came.' This means that the present generation is still fearful of the army," informs teacher Sreemoti.

This is the story of one camp that shows how the temporary army camps, for decades, have made the hill people to panic and have caused damage to the environment. The camp was set up to provide security to the villagers! Why then the camp, even after the peace accord, is still there, is a question the villagers ponder over.

(top) The temporary military camp at Antahapara. (bottom) Supervised by Bangladesh army personnel the members of Shanti Bahini hand over arms at Khagrachhari Stadium. According to the Peace Accord signed in 1997 all temporary army camps should have been closed years back.   © Philip Gain

Stolen FORESTS

Philip Gain

The Chittagong Hill Tracts

Militarization, oppression and the hill tribes

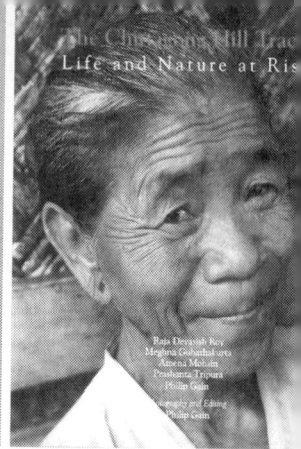

The Chittagong Hill Tracts
Life and Nature at Ris

Raja Devasish Roy
Meghna Guhathakurta
Amena Mohsin
Prashanta Tripura
Philip Gain

Photography and Editing
Philip Gain

The Chittagong Hill Tracts
Living in a Borderland

Willem van Schendel, Wolfgang Mey & Aditya Kumar Dewan

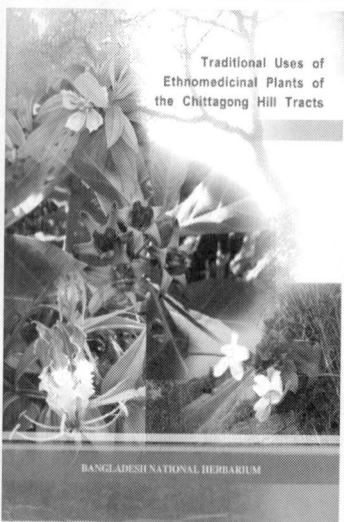

Traditional Uses of
Ethnomedicinal Plants of
the Chittagong Hill Tracts

BANGLADESH NATIONAL HERBARIUM

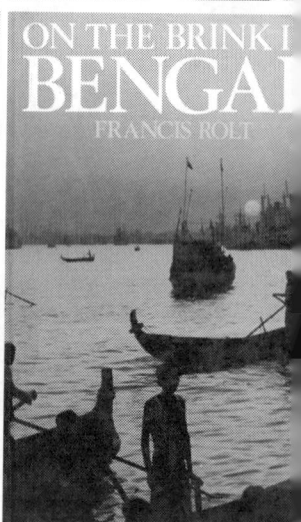

ON THE BRINK I
BENGAI
FRANCIS ROLT

Second Edition

THE Last FORESTS
OF BANGLADESH

Philip Gain

The vanishing old growth forests, mistaken plantations and sufferings of the forest people characterize the forests of Bangladesh.

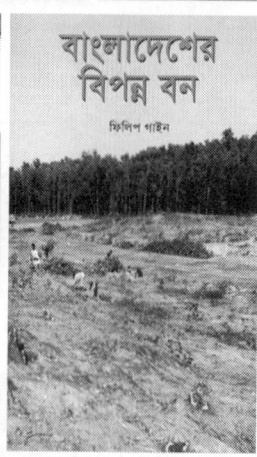

বাংলাদেশের
বিপন্ন বন

ফিলিপ গাইন

বন, বনবিনাশ
ও বনবাসীর
জীবন সংগ্রাম

সম্পাদনা
ফিলিপ গাইন

Critique

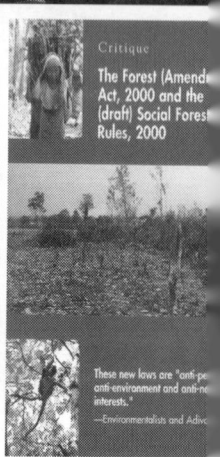

The Forest (Amend
Act, 2000 and the
(draft) Social Forest
Rules, 2000

These new laws are "anti-pe
anti-environment and anti-se
interests."
—Environmentalists and Advo

Resources on the CHT Environment

BOOKS, REPORTS, AND DOCUMENTATY FILMS

The Chittagong Hill Tracts: Life and Nature at Risk

Photography and editing by Philip Gain. 120 pages, 2000. Published by Society for Environment and Human Development (SEHD), Dhaka.

Five contributing writers: Raja Devasish Roy, Amena Mohsin, Meghna Guhathakurta, Prashanta Tripura, and Philip Gain: present information, analyses, photographs, and argument on how the land, life, and nature in the Chittagong Hill Tracts are at risk today. It is no wonder why the region has also witnessed bloodshed, dislocation, disruption, and destruction of life and nature in the name of 'development'. In the nation-state of Pakistan, the hill people were marginalized and remained alienated from the mainstream politics. In Bangladesh their efforts to establish constitutional safeguards: both peacefully and through armed struggle: have also failed. This is a reality full of the agony and anger of the hill people. The peace accord signed in 1997 has ended the bush war and has brought some relief to the region but the amazing, honest, and always smiling hill people are still in a tough struggle to establish their legitimate rights. *The Chittagong Hill Tracts: Life and Nature at Risk* intends to provide basic information on the Chittagong Hill Tracts, stimulate discussion around critical issues, and enhance understanding about the CHT's unique legal and administrative system that has no parallel in other parts of Bangladesh.

Stolen Forests

By Philip Gain. 216 pages, 2006. Published by Society for Environment and Human Development (SEHD), Dhaka.

This is a book of images and critical texts about the dodgy state of the forests of Bangladesh. Author Philip Gain has been through the forests in the hills, coasts, and plains at different times for about two decades. This book reflects his predilection for images of forests; and to him forests are not just trees and the wildlife they support but also the communities that live in the forests, their knowledge, education, history, traditions, technology, culture, and lots more.

The decay of forests is not unique in Bangladesh. But the author makes his point clear that monoculture plantation of teak, rubber, eucalyptus, and acacia have horrendous consequences for the native forests in the Chittagong Hill Tracts (CHT), Cox's Bazar belt, north-central region, and north of Bangladesh. In the coasts, the mangroves patches including the Sundarbans are also faced with massive threats primarily due to prawn aquaculture.

Alongside critical texts, presented are images of the beauty of our native forests and the beautiful faces, minds, and hearts that we see in the forest.

The Chittagong Hill Tracts: Living in a Borderland

By Willem van Schendel, Wolfgang Mey, and Aditya Kumar Dewan. 321 pages, 2000. Published by White Lotus Co. Ltd. Bangkok.

The book is a pictorial introduction to and historical account of the Chittagong Hill Tracts (CHT) of Bangladesh. Covering the time from 1860 to 1970, the book is divided into twenty thematic chapters and a conclusion with four maps and more than 400 mostly unpublished photographs. The book provides its reader with a visual and written account of a region much unknown to the outside world. Using photographs mostly taken by the colonial administrators (both British and Pakistani), tourists, anthropologists, some locals, and the private collections of the elites of the region, the authors have attempted to construct an alternative history of the Chittagong Hill Tracts (CHT). This history focuses on the ruled instead of the rulers, their lives, religions, economies, modes of communications, and cultures.

The book is of immense importance for students of politics, history, and culture. The photographic images forcefully make one aware that Bangladesh is not a land of Bangla speaking people alone. It also demonstrates the hegemonic and exclusivist nature of 'nation' that has denied this space and recognition to the hill people. It makes delightful reading and is an important contribution to the literature on Bangladesh and its pluralist reality (drawn from review by Prof. Amena Mohsin).

Critique: The Forest (Amendment) Act, 2000 and the (draft) Social Forestry Rules, 2000

Editing and photography by Philip Gain. 80 Pages, May 2001. Published by Society for Environment and Human Development (SEHD), Dhaka.

The critique presents critical position and insights of the environmentalists and ethnic communities of Bangladesh on the "Forest (Amendment) Act, 2000" and the "(draft) Social Forestry Rules, 2000". The stated intention of the amendment act passed by the parliament in 2000 and the resulting rules is to promote "social forestry".

But the environmentalists and Adivasis complain that commercial and industrial plantations have been established in the name of "social forestry". They have termed the "Forest (Amendment) Act, 2000" and the "(draft) Social Forestry Rules, 2000" as being "anti-people, anti-environment and anti-national interest" and rejected the new laws. With information, photographs, and reflections, the critique helps one understand the opposition of the environmentalists and ethnic communities.

The Last Forests of Bangladesh

By Philip Gain. 224 pages, 2002. Published by Society for Environment and Human Development (SEHD)

Bangladesh is amazingly green but at the same time a forest-poor land. The country's 18 per cent public forestland has shrunk to approximately six per cent that includes the mangrove forests and the plantations of more than 400,000 ha, raised since 1872. The international financial institutions, ADB and World Bank in particular, financed most of the plantations that are not forest at all.

What factors have led to this situation? The typical response is that growing population, poverty, migration of landless people into the forest areas, shifting cultivation, illegal felling, fuelwood collection, etc. cause degradation of forests. The Forest Department generally blames the people living in and around the forest to be encroachers. But why will the people, materially and spiritually intertwined with the forests, destroy what are so important for their life and environment?

The author focuses on the underlying factors and argues with facts, stories and anecdotes that reserving public forests, plantation of teak, extraction of raw materials from the natural forests for pulp and paper mills, industrial and fuelwood plantation, rubber plantation, so-called "social forestry" and export-oriented prawn cultivation in the recent times are the real causes for the rapid destruction of the forests.

Bon, Bonbinash O Bonobashir Jibon Shangram (Bangla)

(Forest, Its Destruction and Struggle of the Forest People)

Edited by Philip Gain. 263 pages, 2004. Published by Society for Environment and Human Development (SEHD)

The book compiles information and thoughts made at a seminar in June 2003 that was organized by SEHD and write-ups of different authors. With strong research backing, the SEHD researchers and others have illustrated how "simple plantation forestry" mostly with alien and invasive species has ruined the natural forest and the gene pools it used to support in Bangladesh. Plantations of different types particularly on the public forestland, often sugarcoated as "social forestry" have also severely limited the access of the forest-dwelling indigenous peoples to the forests that used to be their commons for centuries. A number of write-ups also explain the lies told about plantation.

Eleven investigative and interpretative reports [of different situations throughout the country], put together in one chapter of the book, show the trend of drastic changes in the forestry industry and unprecedented threats to the forests and forest peoples.

The book also annexes seven declarations that the Adivasis, environmentalists, activists, and development workers have adopted at different times during the last decade. These declarations present aspirations, promises and recommendations that government, external entities and different proponents must pay attention to for the protection of the forest and forest dwelling peoples of Bangladesh.

Parbotto Chattagrame Jumchash (Bangla)
(Swidden Cultivation in the Chittagong Hill Tracts)
By Prashanta Tripura and Abantee Harun. 117 pages, November 2003. Published by Society for Environment and Human Development (SEHD).

This book is a result of the authors' research and discussion concerning *jum* or swidden cultivation in the CHT. Starting from the very definition of swidden agriculture, it contains information about some of the most complicated matters concerning *jum*. The book describes how, from the beginning of the British rule, the *jumias* of the Chittagong Hill Tracts (CHT) started facing opposition and discrimination.

The authors have strong arguments for *jum*. One is, although *jum* cultivation has been the traditional agriculture in the hills for centuries, yet the forests had remained unaffected until recently. It is not the *jum*, but the consistent extraction of raw material for the paper mill, the creation of Kaptai Lake, plantations, illegal logging and timber trade, in-migration of the Bengalis, and militarization that have had devastating effects on the CHT.

Given the general misconceptions and lack of reliable information regarding *jum*, this book will hopefully fill in the information gap to some extent.

Between Ashes and Hope: Chittagong Hill Tracts in the Blind Spot of Bangladeshi Nationalism
Edited by Naeem Mohaiemen. 279 pages, 2010. Published by Drishtipat Writers' Collective, Dhaka.

Between Ashes and Hope is a compilation of contemporary essays, including newspaper op-eds, critical analyses and academic research regarding the burning issues of the Chittagong Hill Tracts. Published in collaboration between Dristipat Writers' Collective and Manusher Jonno Foundation, and edited by Naeem Mohaiemen, this anthology contains contributions from journalists, researchers, members of indigenous communities and the civil society. Translations of various published Bengali and Jumma language essays into English point to the all encompassing approach to integrate diverse accounts. While the subject of the CHT region is largely suppressed or even fabricated by the mainstream Bangladeshi media, the book attempts to bring forth the post liberation exploitation, settlement, indifference and colonization by the Bangladeshi regimes, and stresses the importance of implementation of the 1997 Peace Accord.

Genocide in the Chittagong Hill Tracts, Bangladesh

Edited by Wolfgang Mey. 193 pages, 1984. Published by International Work Group for Indigenous Affairs, Copenhagen.

This report relates primarily to the abuses including genocide in the Chittagong Hill Tracts (CHT) of Bangladesh. The report presents stark facts about state-sponsored Bengali [poor and landless from the plains districts] settlements in the Chittagong Hill Tracts in the 1980s and subsequent violence including evictions, rape and massacres against the indigenous peoples. To put the picture of atrocities in the words of the editor of the report, "They have been subjected to large scale evictions torture, rape and massacres--10,000 tribal people were killed in 1981 alone." The perpetrators: Bengali settlers and the military: committed such crimes to drive away the indigenous peoples from their ancestors' land and to contain an insurgency war that was triggered by politically-motivated in-migration. The report also provides a solid background on historical changes that help understand the CHT and its peoples. The report also contains maps and photos on the location of ethnic groups and the surroundings of the CHT. Those looking back into the times of upheavals during the insurgency in the CHT will find the report very valuable.

The Chittagong Hill Tracts: Militarization, oppression and the hill tribes

96 pages, 1984. Published by Anti-Slavery Society, London.

It's a landmark publication by the Britain based Anti-Slavery Society (now Anti-Slavery International), predominantly voicing the much repressed and obscured account of the indigenous people who view themselves victims of "policy of militarization and internal colonization" by the administration. Published in 1984, it provides a valuable commentary on the tribal cause in Bangladesh throughout history, from their stable status guarded by the Chittagong Hill Tracts 1900 Regulation during the British colonial period, to their subsequent degradation during the Pakistani regime and under the post-liberation military doctrine, which further intensified their oppression. The book also resonates their mass discontent with the various government development initiatives such as the Kaptai Dam, the Chandraghona Paper Mill, as well as non-government programs of the Swedish International Development Authority (SIDA) and the Austrian Development Assistance Bureau (ADAB), which they view as tools of exploitation and deprivation. Supported by excerpts from various international publications as well as concealed letters and memorandum, it reveals detailed accounts of the atrocious massacres committed by the military against the resistance movement in the name of nationalism. The book is also an insightful documentation, justifying the rise of the controversial *Shanti Bahini* of PCJSS and substantiating their preferred social and political order apart from the policies of 'the tyrannical government in Dhaka', the perspectives, which transpire to be prevalent till today.

The Charge of Genocide: Human Rights in the Chittagong Hill Tracts of Bangladesh

222 pages, October 1986. Published by Organizing Committee Chittagong Hill Tracts Campaign, Amsterdam.

This publication is a compilation of papers, reports, and documents presented at a conference on armed conflict and human rights violations in the Chittagong Hill Tracts (CHT). The papers assembled into eight chapters deal with the historical roots of conflict, militarization, population dislocation, in-migration, local resistance and international protests against human rights abuses, the role of foreign aid and the so-called economic development that militate against the indigenous communities, and why the world should care about the CHT. The papers also give a rough understanding of how the situation in the CHT has progressed from the British rule up until the point of this publication. Different chapters contain government correspondence including secret memos, information published in the print media, expert analysis and views on human rights violations and genocides. One looking back into the times of armed conflict in the CHT will find this publication resourceful.

Francis Buchanan in Southeast Bengal (1798): His Journey to Chittagong, The Chittagong Hill Tracts, Noakhali and Comilla

Edited by Willem Van Schendel. 209 pages, 2009. Published by University Press Ltd. Bangla edition (214 pages) is also available.

The development history of Bangladesh is marked by fluctuations, turn-abouts and decelerations. To understand the complex dynamics of social and economic change towards finding solutions, we need to dig for historical knowledge. Francis Buchanan's account is the earliest and the most significant source of new information on the eighteenth Century Bengal, Arakan, Tripura, Cachar, Mizoram and Burma (today's Myanmar). It provides unique information on Southeastern Bengal in particular and further regions to the South and the East in general and is a good example of how Europeans collected knowledge of the wider world and what views they held. It contains information on rural economy, social life, and ethnic relations and above all, the British imperial policy in the region. Although Buchanan's account is presented as a travel diary, it is the diary of a disciplined traveler. He is the first visitor to the Chittagong Hill Tracts to have left a careful record in the form of a diary. His information is simply invaluable for the reconstruction of the history of this area. He describes the situation sixty years before the British annexation (1860); the first accounts of the Chittagong Hill Tracts that were known so far date from after that annexation. His information is extraordinarily rich and well researched. He combines his own observations with interviews with a considerable number of Marma, Chakma, Mru, Zo, Tippera, Bengali, Burmese, and Arakanese informants.

On the Brink in Bengal

By Francis Rolt. 181 pages, 1991. Published by John Murray
Ltd., London.

A travelogue and descriptive account, *On the Brink in Bengal,* takes us on
a fascinating trail down the border region of Bangladesh, where smaller
ethnic communities and the majority Bengalis have mixed, met, fought
and traded for centuries. Combining acute observation with humor, and
listening to what everyone he met had to say, Francis Rolt gives us an
illuminating insight into the lives and thoughts of peoples on the margins
of life in the great sub-continent. Francis Rolt managed to elicit myths,
stories and tales from those he met during his journey through "tribal"
villages along the eastern border of Bangladesh. Stranded on a tropical
island in the Bay of Bengal, detained by the army in the forbidden jungles
of the Chittagong Hill Tracts, arguing with a snake charmer, Francis Rolt
and photographer Peter Barker got themselves into and out of a surprising
number of scrapes. The border region of Bangladesh is densely populated
and confusing. On one side are "tribal" people, different in every imaginable
way from the people of plainsland Bengal, and the tension produced by the
dramatic, historical clash between these two great cultural blocks is the
starting point for the journey.

Bangladesher Biponno Bon (Bangla)

By Philip Gain. 276 Pages, 2005. Published by Society for
Environment and Human Development (SEHD), Dhaka.

Bangladesh is amazingly green with myriad biodiversity resources.
But at the same time it is a forest-poor country that has lost its forest
cover from about 20% in 1927 to a mere six per cent today. Outside the
Sundarbans, only tiny patches of forests survive today. What factors have
led to this perilous condition? The typical response that comes is growing
population, poverty, migration of landless people into the forest areas,
shifting cultivation, illegal felling, fuelwood collection, etc.

The underlying factors for the destruction of the forests are an area where
Philip Gain has investigated for the last one and a half decades. *Bangladesher
Biponno Bon* (Endangered Forests of Bangladesh) is the outcome of his
investigation. It's an updated Bangla edition of his book, *The Last Forests
of Bangladesh.* He is not a forestry professional. He writes this book due

to his deep interest in environment, human rights and Adivasis. Anyone with genuine interest in environment, forests, and the Adivasis will benefit from this book.

Chittagong Hill Tracts Best Practices Handbook

Published by United Nations Development Program (UNDP), 2005. Dhaka, Bangladesh.

This handbook recognizes and award tributes to the people of the CHT in shaping their own future, sometimes under inconvenient circumstances and with limited resources. This book demonstrates the remarkable creativity, innovation and pragmatism of the people of the CHT how they find practical solutions to everyday challenges for the betterment of their lives. The practice book dedicates to sharing their 'development story' over these many years and spreads their many models of 'best practices' and guides to an inspiration for self-reliance in the future. In recording and sharing this story, the Chittagong Hill Tracts Development Facility (CHTDF) of UNDP promised in collaboration with FAO, ILO, UNESCO, UNFPA and WFA who develop a Resource Database & Directory (RDD) by exploring, identifying and recording the 'best practices (BP)'. The RDD exercise includes a vast dimension of sectors like Agro-forestry, Crops, Fisheries, Livestock, Environment, Health and Family Planning, Education, Enterprise Development, Employment Creation and the small infrastructure of CHT. It carries out with the intention of enabling CHT communities, Government institutions and NGOs to identify promising development options and to facilitate local decision making in the CHT.

Chakma Talik Chikitsa (Bangla)
(Chakma pharmacopeias)

by Dr. Bhagadatta Khisa. 136 pages, 1996. Moni Swapan Dewan, Rangamati

There are hundreds of books about the history, heritage, culture, and language of the Chakmas of the CHT but none about their traditional medicinal practice. The only exception is *Chakma Talik Chikitsa* (Chakma pharmacopeia). In this book, the writer describes different diseases with their different names according to the harmful deities responsible for them. Then he describes the symptoms of each disease, the procedure to make the medicines, and how to apply them. He mentions four categories of medicines that are used in Chakma *talik* and one of them is herbal medicine. He also gives lists of the herbal plants and other ingredients to make the medicines. The names based on the deities and the spells to cast them out could be ignored as those are unscientific and superstitious. However, the list of diseases, symptoms, and cures could be very helpful both for the patients and the researchers.

Chittagong Hill Tracts: State of Environment

Edited by Quamrul Islam Chowdhury. Pages 147, 2001.
Forum of Environmental Journalists of Bangladesh (FEJB.
Dhaka.

The book highlights the ecological problems of the CHT, the southern part of Bangladesh that is a rugged, hilly and jungle clad region. Quamrul Islam Chowdhury, the editor of the book admits that the book is not a systematic study of the problems of the CHT. However, the book focuses on the exotic biodiversity, the hills and the jungles, and the diverse ethnic groups who enrich the country's national culture and heritage. Different topics involving the CHT are discussed in the articles and reports compiled in the book such as the environmental plans for CHT, the decline of the village common forests (VCFs), physical features of the region, Sustainable Environmental Management Plan (SEMP), the rights of the indigenous people, their culture and history, and natural disasters like landslides.

Socio-economic Baseline Survey of Chittagong Hill Tracts

By Abul Barkat. 307 pages, 2009. Published by United
Nations Development Program, Dhaka.

As it is evident from the name, this book is the outcome of a baseline survey of the Chittagong Hill Tracts, which is the home to 12 indigenous communities of Bangladesh who are one of the most disadvantaged and vulnerable in terms of various developments indicators, including access to and ownership of land, income, employment opportunities, poverty, housing, health, water, sanitation, education, and inter-community confidence. The survey, aimed at generating benchmark information, provides data that help one to understand the socio-economic status of the people living in the Chittagong Hill Tracts.

Counting the Hills: Assessing Development in Chittagong Hill Tracts

Edited by Mohammad Rafi and A. Mushdtaque R. Chowdhury. 261 pages, 2001. Published by The University Press Limited, Dhaka.

This book is a formative and statistical analysis on the CHT through the use of field experience as well as conceptual exploration. It is comprised of demographical, social, economic, educational, health, and environmental research of the five major indigenous groups of the CHT: Chakma, Marma, Mro, and Tripura: and Bengalis, based on population concentration, by BRAC. The authors use a survey, as well as a cross-comparison of the relative status of the individual groups based on their opinions of one another. In this way they provide a qualitative as well as quantitative assessment of these major groups. Furthermore they use this analysis to draw conclusion on what may not be only the most effective, but also the most efficient avenues for policy implementation regarding the development of the CHT. The authors also inform the readers as to other actions that could be taken (i.e. the complete implementation of the Peace Accord of 1997) in order to facilitate the development of the region.

Migration, Land Alienation and Ethnic Conflict

Causes of Poverty in the Chittagong Hill Tracts of Bangladesh

By Shapan Adnan, 349 pages, 2004. Published by Research and Advisory Services, Dhaka.

This is the outcome of a study of the causes of poverty among the indigenous peoples of the Chittagong Hill Tracts (CHT). It deals with the social, economic and demographic aspects of that region. However, the focus is on the causes of poverty of the Hill peoples popularly known also as Jumma people or the Paharis. The book identifies four broad types of mechanism generating and sustaining poverty among the Hill peoples. These are manifested in a large number of processes, including exploitation and surplus extraction, expropriation of land and property, loss of lands, forests, and biodiversity, constraints to economic growth and human development, as well as breakdown of community-based institutions and redistributive norms among the Hill peoples undermining their erstwhile capability to prevent or to cope with poverty. These poverty-generating processes, in turn, are found to have been driven by antecedent factors and forces. The concluding chapter of the volume puts forward policy recommendations to deal with the manifold causes of poverty among the hill peoples, inclusive of land alienation and ethnic conflict in the CHT. The book is enriched with nearly 200 color plates illustrating various sites and contexts of the region and the hill peoples.

Parbotto Chattagram Ain Songhita (Bangla)
(Chittagong Hill Tracts Law Compendium)
Published by CHT Regional Council. 570 pages, 2010. with assistance of United Nations Development Programme-Chittagong Hill Tracts Development Facility (UNDP-CHTDF)

The book is basically a compilation of all laws, regulation, rules, orders and gazettes that apply to the Chittaong Hill Tracts. It also compiles laws that have limited application and the laws annulled. The compendium has five parts. The first part contains unique laws formulated in the light of the peace accord signed in 1997; different traditional and special laws and rules formulated in the light of rules under these special laws; and several laws [including interpretation] relating to local councils. The second part compiles several laws and documents [including interpretation] relating to the Chittagong Hill Tracts. The third part gives a list of several noted general laws and rules applicable in the light of the CHT regulations. The fourth part contains documents based on which this law compendium has been prepared. The fifth part is the index.

Bangladesh District Gazetteers: Chittagong Hill Tracts
General Editor. MUHAMMAD ISHAQ. 319 pages, 1975.
Published by Bangladesh Press.

It's an important document that was first published during the British colonial times (in the early twentieth century). With 14 chapters it gives accounts on physical aspects, history, people, society, culture, agriculture, livestock, forests, economic condition, communications, industries, trade and commerce, public health, education, language and literature, administration (general, land, and local), local government system, and places of interest in the Chittagong Hill Tracts. Many changes have happened [in administration and politics] in the CHT since this document was last published in 1975. But what is unique of this document is that it gives accounts that help one to understand how unique the CHT environment was, the region's natural history, and what Bangladesh has lost. The document is no more available in the market [although the gazetteers on other districts are still sold]. There is alleged restriction from certain quarters in the government on the sale and distribution of this document.

Chittagong Hill Tracts: Soil and Land Use Survey 1964-1966

Forestal Forestry and Engineering International Limited, Vancouver, Canada. 1 to 9 Volume, 1962. Canadian Colombo Plan, East Pakistan Agricultural Development Cooperation

Three years after implementation of the Kaptai hydro-electric project the Canadian company, Forestal Forestry and Engineering International Limited was commissioned to study the possible impacts of the dam on the ethnic people and the future developments in the Hill Tracts. "The tribal people had attained a reasonably satisfactory way of life adequately adjusted to the limitations imposed by the physical environment" was a key finding of Forestal (Anti-Slavery Society's report, "The Chittagong Hill Tracts: Militarization, oppression and the hill tribes, 1984: 33). "After the dislocation, however, the company reported that a disastrous cycle of over-cultivation had led to depletion of soil fertility, loss of forest cover, serious erosion and further increased pressure on the remaining land." (Anti-Slavery Society 1984:33). The report done by an 11-member team of geologists, economists, agronomists, and biologists strongly criticized *jum* or shifting cultivation. A key recommendation of the team was that shifting cultivation be controlled and horticultural production be introduced. The anti-*jum* sentiment is rooted in this report. What the team proposed as a result of their research (Forestal Report 1966) "has formed the basis of most of the government plans for development of the Chittagong Hill Tracts in the following years."

The nine-volume Forestal Report is worth consulting in understanding the then Pakistani State's attitude and conviction about the CHT, its peoples, and traditional agriculture that have had far-reaching effects on environment and life of the indigenous people of the region.

Traditional Uses of Ethnomedicinal Plants of the Chittagong Hill Tracts

By Sarder Nasir Uddin, Edited by Dr. M. Matiur Rahman. 891 Pages, 2006. Published by Bangladesh National Harbarium.

This book is an outcome of a project formulated by the Bangladesh National Herbarium and sponsored by the Ministry of Chittagong Hill Tracts Affairs. Medicinal plants constitute an important natural wealth of Bangladesh especially in the three hill districts of Khagrachari, Rangamati, and Bandarban. Herbal Medicine prepared by the herbal practioners (Boidyas) has been the only source of treatment of the indigenous peoples of the CHT for their primary health problems. In the text of the book the chapter, "Tribal Prescription" includes the name of 302 diseases and ailments and enunciates their treatments with 2,295 prescriptions by the herbal practioners with 700 vascular plant species. All these communities have had rich tradition, culture, belief and herbal heritage. The data presented were recorded after proper scrutiny and repeated verifications both in the field as well as in the National Herbarium.

Medicinal Knowledge and Plants of Chittagong Hill Tracts, Bangladesh

By Khondkar Ismail. 420 pages. Unpublished.

This book is focused especially on indigenous health treatment by using traditional knowledge. The indigenous people, for generations, have acquired a vast heritage of knowledge and experience of traditional treatment. Indigenous peoples use their secret knowledge for health treatment. They follow "Indigenous Pharmacopeias" for treatment. They never hand it over to others. The author discovered this book, gathered information from it about various herbal plants, and carried out further research in the USA and Australia to write his own book. In the hill areas, the indigenous communities use this book for their health treatment. The book of 420 pages contains names (in Chakma, Marma, and Tanchangya languages) and details of 100 medicinal plants including some original parts of Pharmacopeias. (For further details see a review of the book by Mosabber Hossain, *Prothom Alo*, 7 September 2012; the book was yet to be published as of Sept. 2012).

Departed Melody (Memoirs)

By Raja Tridiv Roy. 456 page, July 2003. Published by PPA Publication, Islamabad.

This book is an autobiographical recounting of the life of Raja Tridiv Roy, the former chief of the Chakma Circle who has lived his life as a Pakistani citizen since Bangladesh became independent in 1971. The book presents an interesting take on the modern history of the Chittagong Hill Tracts, the events surrounding the war between East and West Pakistan, and the life of a Chakma-cum-Pakistani elite. The book also contains information and insights that help understand the environmental issues of the CHT.

Though much of the historical information presented is well known, what the book offers is a bit of insight into the complexities surrounding the Hill Tracts and the Liberation War, which fall outside of the ambit of the nationalist history. One must read between the lines and around them to get any insight into Roy's rationale as someone so handsomely rewarded for his cooperation with the Pakistani cause.

Documentary Films

Teardrops of Karnaphuli (Kanaphulir Kanna)

Produced and by KINO-EYE FILMS. Script and direction by Tanvir Mokammel (T: 880-2-8127638, E: tmokammel@yahoo.com, I:www.tanvirmokammel.com

A 60-minute documentary film on the effects of development projects in the CHT, the state-sponsored Bengali settlements on the land that has been the traditional *jum* land of the indigenous peoples for centuries, massive scale human rights violations, armed conflicts, and human miseries suffered mainly by the hill peoples.

Biponno Bon (Stolen Forests)

Produced by Society for Environment and Human Development (SEHD). Concept, research and screenplay: Philip Gain, Direction: Philip Gain and Junaid Halim, English 76-minute 1998, Bangla 45-minute 1998 (available in DVD and CD)

A 45-minute documentary film on the forests of Bangladesh (except for the Sundarbans) that have been devastated. The hills in the Chittagong Hill Tracts are bare today. The traditional *sal* forest has become history in most parts. The monoculture plantations of exotic and invasive species in place of hundreds of species of the native forests are not forests at all. One key message of the film is man can plant forest but cannot create it.

Khumi Lives

Produced by Hill Film Society. Research and Direction by Ittukgula Changma. Camera work by Ittukgula Changma, Anutosh Changma, Joyanta Howlader, Wahid Ashraf Rony.

A 32-minute documentary film on the daily life, language, rituals, myths, occupations, religions, and other social matters of the Khumis. The Khumis, a pure *jumia* (swidden cultivators) indigenous community of the CHT with a population of around 2,500, has been living in the Chittagong Hill Tracts for a long time. The film reflects on the life cycle of the Khumis. Traditional *jum* or swidden cultivation is at the center of the film.

Maleya Taroom

Produced by Tamaza for Taungya. Script by Dvasish Wangza. Photography by Amiyo Kanti Chakma, Suvasish Chakma, Devasish Wangza, and Subrata Chakma.

A 15-minute documentary film tells the story of the village common forests (VCFs), an old-day method to retain forest cover around the settlements of the indigenous peoples, in the Chittagong Hill Tracts (CHT). The film also documents the efforts of Taungya, a CHT-based organization in promoting VCFs.

Review by: Philip Gain, Alimul Haque, Robert Alec Lindeman, Asif Khan, Tania Sultana.

Glossary, Concepts, and Theories

Adivasi: indigenous or aboriginal (Bangla/Hindi). In Bangladesh Adivasis are also known as `ethnic communities' and `tribal'. According to government census there are 27 'tribal' communities and they account for 1,410,169 (592,977 in the CHT and 817,192 in the plains) or 1.13% of the country's total population (2001). The actual number of ethnic communities may be as many as 45 and their population is also believed to be greater than the government-produced figures. Eleven of the ethnic communities live in the CHT and the rest are concentrated in the Northwest, North-central and Northeastern regions of Bangladesh.

Acacia (mangium): A tree species native to the islands of Sula, Ceram and Aru in Indonesia, the western province of Papua New Guinea and northeastern Queensland, Australia. It is a major plantation species in Asia. In recent times, the tree has been planted in large numbers in Bangladesh as a mono-crop on public forestland under different plantation projects. The tree is seen growing abundantly in the CHT and its outskirts.

Agroforestry: An agricultural practice in which trees or shrubs are grown alongside food or other crops. Agroforestry on public forestland in Bangladesh is similar to woodlot that is intended for the production of fuelwood.

Alien species: A non-native life form that has evolved in one location and travels to or is introduced to an area where it does not naturally occur. Many alien species are not able to establish themselves in their new environments. These alien species that do so are called invasive species.

Animism: Animism (from Latin anima, meaning "spirit", "soul", "life force") is a worldview that believes that life force is omnipresent, most evidently in the living things–trees, plants, birds, animals, and human beings, but also in inanimate conditions such as the motion of water, the sun, the moon, clouds, wind, static mountains and soil. One of man's oldest surviving beliefs, rooted in philosophy and religion; it is often associated with oral religions whose followers attribute sanctity to the inanimate things as to the spirit of humans.

Animism might have generated from the primitive notion that spirits and souls, after death, pass from one body to another in the form of the shadow of human beings, as well as of the inanimate objects, plants, and animals. Thus, it emphasizes respecting and understanding the reality and conduct of the all-encompassing spirit world, to garner blessing and to avoid harm.

"The first peace, which is the most important, is that which comes within the souls of people when they realize their relationship, their oneness, with the universe and all its powers, and when they realize that at the center of the universe dwells Wakan-Taka (the Great Spirit), and that this center is really everywhere, it is within each of us.": Black Elk-Oglala Sioux.

Bangladesh Forest Industries Development Corporation (BFIDC): A corporation established in 1959 to extract timber/rubber wood and other forest produces from forestland and BFIDC's rubber garden. Its another main objective is to establish industries and factories for commercial use of forest produces and rubber wood.

Biodiversity: Biological diversity. The diversity of living organisms on earth.

Complex plantation: A plantation system that is structurally more complex, yields a wider range of forest products, supports greater species diversity and is more integrated with the needs of local populations, than a simple plantation system.

Dao: hewing knife, the most widely used work tool; bamboo or wooden shaft with inserted iron blade; shaft length ca. 20 cm. [8 inch], blade (of various forms) ca. 35 cm (14 in.) (Brauns and Löffler 1990: 70)

Deforestation: The clearing of forests and conversion of the forestland to non-forest uses.

Degraded forest: A forest that has lost the productive potential of its natural resources due to biological, chemical and physical processes. A degraded forest cannot support the same species diversity and population size as a natural, undisturbed forest can.

Deep ecology: Term used by Norwegian philosopher Arne Naess to signify that humans are a part of nature, a strand in the web of life; that nature is a cohesive whole rather than a sum of parts; and that economic goals should be subordinate to ecological concerns. This term is used in contrast to shallow ecology, which sees humans as ruling nature, analyzing and breaking nature into understandable parts, and puts economic goals ahead of environmental concerns.

Ecosystem: Plants, animals, and other life forms in a defined area, interacting with each other in their physical environment.

Ecocide: Widespread destruction, damage to, or loss of ecosystem(s) of a given territory, whether by man or by other causes, to such an extent that peaceful existence by the inhabitants of that territory has been severely reduced.

Ecotourism: Ecotourism involves visiting fragile, pristine, and relatively undisturbed natural areas, intended as a low-impact and often small-scale alternative to standard commercial (mass) tourism. Its purpose is to educate the traveler, to provide funds for ecological conservation, to directly benefit the economic development and political empowerment of local communities, or to foster respect for different cultures and for human rights.

Environmental justice: Environmental justice is the fair treatment and meaningful involvement of all people regardless of race, color, national origin, or income with respect to the development, implementation, and enforcement of environmental laws, regulations, and policies. In linking environmental and social justice issues the environmental justice approach seeks to challenge the abuse of power, which results in poor people having to suffer the effects of environmental damage caused by the greed of others.

Ecology: Ecology is the scientific study of the relationships that living organisms have with each other and with their natural environment. Ecology is an interdisciplinary field that includes biology and earth science.

Ecology is a human science as well. There are many practical applications of ecology in the conservation of biology, wetland management, natural resource management (agro-ecology, agriculture, forestry, agro-forestry, and fisheries), city planning (urban ecology), community health, economics, basic and applied science, and human social interaction (human ecology).

Eucalyptus: A tall tree with a characteristic smooth, flaky, white-ash bark and fragrant leaves. A tree native to Australia, it is a fast growing species and an established exotic in Asia, Africa and Latin America utilized for pulpwood, oils and medicines. It is controversial because of its extensive plantation and it not providing fodder, foliage or fruit like the native trees it replaces. The eucalyptus tree also consumes massive amounts of water, which depletes both soil moisture and groundwater reserves.

Evergreens: Trees that retain living leaves throughout the year. The leaves of a previous season are not dropped until the growth of the new foliage is complete.

Forest dependent communities: Communities that depend largely on forests for their livelihood. In Bangladesh, forest dependent communities are mostly ethnic communities, which live in the CHT and other parts of the country. In the CHT in particular, communities, which still depend on *jum* for their subsistence are pretty much considered to be forest communities.

Fallow: Previously cultivated land that is left unseeded and uncultivated for a period of time in order to allow vegetation to regenerate, and the soil to regain nutrients that have been depleted by crops.

Headman: Head of a *mouza* [in the CHT] charged with the collection of revenue, land, "tribal" and justice administration at the *mouza* level. He also supervises the work of the *karbaries* and is responsible to the chief of the circle and the Deputy Commissioner (DC). Ideally, he is supposed to play an important role in the management of the land and forests, but given the ground realities in the CHT, he is often ignored by the civil administration officials.

Herbal medicine and treatment: Herbal medicine, the botanical medicine refers to using plants, seeds, berries, roots, leaves, barks, or flowers for medicinal purposes. The scope of herbal medicine is sometimes extended to include fungal and bee products, as well as minerals, shells and certain animal parts. It is used for the treatment of various diseases. Many people in remote areas of Bangladesh do not get the health care of the modern medical system. They have vast knowledge of use of plants as a source of medicine. Some *boiddayas*, the elderly men and women, give treatment with the herbal medicine in remote areas.

Homestead forest: Different kinds of trees grown around homes in rural areas meant to provide fuelwood, home construction materials, fruits, etc.

Industrial plantation: Stand of planted trees of the same age and usually of the same species for the production of industrial forest products such as sawlogs, veneer logs, pulpwood and poles. Industrial plantations require extensive and repeated human intervention and do not act as a habitat for local species or as a resource for local communities.

Internal displacement persons: An internally displaced person (IDP) is someone who is forced to flee his or her home but who remains within his or her country's borders. They are often referred to as refugees, although they do not fall within the current legal definition of a refugee. Internal displacement is a big issue in the CHT related to environmental disasters. In the CHT, the origin of internal displacement is rooted in the Kaptai Hydroelectricity project and construction of a huge dam that necessitated displacement of a fourth of the CHT population (more than 100,000) in the 1960s. Then the insurgency war that the hill people fought with the Bangladesh army caused another wave of internal displacement. According to Human Development Research Centre (BHDRC) between 1997 and 2007 there are as many as 323,000 IDPs in the CHT (BHDRC, April 2009).

Invasive species: Alien species that successfully establish themselves in the environments in which they have been introduced. Invasive species often endanger the survival of native species by preying on them, competing with them for limiting resources and altering the native habitat so as to make survival difficult for native species.

Jote and jote permit: *Jote* refers to the document of the land owned by any individual. 'Jote Permit' is an official permission issued by the FD to fell trees from a particular *jote.*

Jum cultivation (also shifting, swidden or slash and burn cultivation): A traditional agricultural system practised in the Chittagong Hill Tracts. It involves the clearing of jungle and the burning of dried debris, followed by a year or two of cultivation in a *jum* field and then a fallow period of several years in which vegetation is allowed to reemerge. On the mountain slopes *jum* cultivation is the only method of growing crops without causing too much detriment to the soil.

Jumia: *jum* cultivator. In the past most of the ethnic communities in the CHT were *jumias*. Despite constant government efforts to discourage *jum* cultivation, a large percentage of hill ethnic people remain *jumia*, having no other option left for subsistence.

Karbari: Hereditary head of a hamlet in the CHT, traditionally nominated by the villagers, formally appointed by the circle chiefs, and acknowledged by the administration. A *karbari*, with no official power or authority, is responsible to deliver the field taxes to the headman. S/he receives no remuneration for services rendered.

Medicinal plant: Plant used for treatment of numerous health problems. The mainstream pharmaceuticals companies use it in preparing herbal medicine. Medicinal plants constitute an important natural wealth of any country. They play a significant role in providing primary health care services to rural people. A substantial amount of foreign exchange can be earned by exporting medicinal plants to other countries. In this way indigenous medicinal plants play a significant role in the economy of a country. Bangladesh, a tropical country, has hundreds of medicinal plants in use.

MNC (multinational company or corporation): A corporation or enterprise that manages production establishments or delivers services in at least two countries. Very large multinationals have budgets that exceed national budgets of many countries. Multinational corporations can have a powerful influence in international relations and local economies. Multinational corporations play an important role in globalization.

Monoculture: The practice of raising a single species, generally even-aged, in a plantation as opposed to the large number of species of trees found in a native forest or in a mixed plantation.

Mouza: The smallest regional unit of fiscal administration in the Chittagong Hill Tracts, under the authority of a headman. A *mouza* is made up of several small, independent villages (hamlets) that are administratively united irrespective of ethnic composition. There are 376 *mouzas* in the CHT.

Native species: Original inhabitant of an area, region or country. A species that has evolved in a specific environment is native to that place.

Plantation: Commercial crop of trees of one or a very few (often alien) species. Plantations require extensive human interference to prepare soil, maintain trees and harvest trees. Trees are planted in homogenous blocks of the same age. Plantations are devoid of the diverse flora and fauna present in native forests. Access to plantations by local communities is generally restricted.

Pulp: A substance within wood used for the manufacture of paper. It is produced by grinding wood (mechanical pulp) or chipping it and boiling the chips with chemicals before refining it (chemical pulp). One ton of bleached chemical pulp requires 20 plantation trees or 4.8 cubic meters of wood, 120,000 or more liters of water and 1.2-megawatt hours of electricity (Carrere and Lohmann, 1996: 19). The production of one ton of bleached pulp also requires one to three kilograms of sulphur based chemicals, which produce sulphur dioxide, 50–80 kilograms of chlorine, which reacts with chemicals in pulp to produce hundreds of organochlorine pollutants, and it releases toxic mineral salts such as aluminium (Carrere and Lohmann, 1996: 83).

Pulpwood plantation: Plantation intended to produce wood for making pulp. Industrial plantation is also often intended for the production of pulpwood. (Add CHT situation).

Raja (also circle chief): An administrative head of a circle. There are three circles in the CHT: Chakma, Bohmong, and Mong. The *rajas* or chiefs are responsible for the administration of `tribal' justice and the customary laws of the hill people, revenue administration and advising different civil administrations.

Rubber: The processed latex of the rubber tree, *Hevea brasiliensis*. It is collected by making an incision in the bark of the tree and allowing it to drip into cups. The latex is then processed. The rubber tree is native to Brazil. It is now found in vast rubber plantations throughout Southeast Asia including Bangladesh.

Second-growth forest: The re-emergence of trees, vegetation and fauna following a significant disturbance of the original environment. The new forest often differs in both structure and species composition from the primary forest.

Simple plantation: A drastically simplified forest system intensively managed to produce various tree products such as resin, oil and lumber. Trees in simple plantations are generally fast growing species that enable a greater number of wood harvests than natural forests. (Woodwell 2001: 104). A simple plantation is the graveyard of the native forest and all the life forms it supports.

Social forestry: People-oriented forestry policy and practises that aim at ensuring active participation by the rural individuals and communities in planning, implementation and benefit sharing in forestry and tree-growing activities.

Sustainable development: Development, such as economic or social, that meets the needs of the present without compromising the ability of future generations to meet their own needs.

Teak: *Tectona grandees,* a large deciduous tree, is not indigenous to Bangladesh. It is one of the most valuable timber species. It has straight, even bore and small white flowers. Teak, planted in the CHT in large areas, however, has proven to be ecologically unsound especially when planted as monoculture in a patch. It taxes the soil too much. Teak, a deciduous species, increases fire hazards and is susceptible to attack by teak defoliators and the teak canker insects.

Tea: Tea, the second most popular beverage in the world (the first is water) is believed to have been first popularized in China. It is made from the young leaves and leaf buds of the tea plant, *Camellia sinensis.* Two principal varieties in use are: the small-leaved China plant (*C. sinensis sinensis)* and the large-leaved Assam plant (*C. sinensis assamica*). Tea was discovered growing wild in Nepal and Monipur in 1820.

Timber: A large squared or dressed piece of wood ready for use or forming part of a structure. It has a universal human and commercial value. It was not long ago when the forests in the Chittagong Hill Tracts had myriad species of timber trees harvested by the state authorities for even supply of railway sleepers. The local communities used to freely harvest timber for construction of their dwellings and canoes among other things. They still do harvest timber, mainly from their own gardens. However, nowadays the timber trees in the CHT have reduced to such point that it is no more harvested officially.

Unclassed state forest (USF): Three-fourths of the CHT is USF. The ethnic peoples of the CHT practice their traditional *jum* cultivation mostly in the USF that is targeted for the promotion of plantation economy. Land given to government agencies, private parties, companies, outside individuals and the local elite for rubber cultivation, production of pulpwood and other commercial purposes belong chiefly to this category.

Woodlot: Plantation, simple in most cases, for the production of fuelwood.

Nappi: Also "sidol", an essential curry paste in the hill peoples' kitchen made from fish or shrimp. The Marmas are particularly skilled in making *nappi*.

Political ecology: is the study of the relationships between political, economic, and social factors with environmental issues and changes.

Tobacco: *Nicotiana tabacum* native to Brazil is named after Jean Nicot, French Ambassador to the Portuguese Court at Lisbon. There are sixty-six species of the tobacco genus of which only one, *Nicotiana tabacum,* is intended for smoking, sniffing, chewing or extraction of nicotine. The others, except for those grown for their handsome flowers, are to be avoided due to their high toxicity.

Union Parishad (UP): Union Parishad (Union council) is the smallest rural administrative and local government unit in Bangladesh. There are 4,451 UPs in Bangladesh. Each UP consists of a chairman and twelve members including three seats exclusively reserved for women. The primary responsibility of the UP is to develop agriculture, industry as well as community within the local limits of the union. The chairman has the power to conduct trials and give punishments for certain minor crimes.

Upazila: sub-district. Of 486 upazilas in 64 districts of Bangladesh, the Chittagong Hill Tracts has 25. Upazilas are similar to the country sub-divisions found in some Western countries. The upazilas are the second lowest tier of regional administration in Bangladesh.

Village Common Forest (VCF): Forests managed by the indigenous village communities in common interest and for ecological reasons.

References

"Debt bondage", Wikipedia® is a registered trademark of the Wikimedia Foundation, Inc. This page was last modified on 13 October 2010 at 19:10.

"Multinational corporation", Babylon Translation. Copyright © 1997-2009 Babylon Ltd. All Rights Reserved to Babylon Translation Software. <http://dictionary.babylon.com/multinational_companies/> Oct. 14, 2010.

"Peace", Dictionary.com, LLC. Copyright © 2010. All rights reserved. <http://dictionary.reference.com/browse/peace> Oct. 14, 2010.

<http://en.wikipedia.org/wiki/Debt_bondage> Oct. 14, 2010.

Brown, Graham K.; Beneulin, Séverine; & Devine, Joseph. August 2009. *Contesting the boundaries of religion in social mobilization*. Center for Development Studies, Bath University.

Brundtland Commission. *Our Common Future*. Published by Oxford Press 1987.

Calcutta Research Group. 2007. *Rethinking Rights, Justice, and Development*.

Carrere, Ricardo and Lohmann, Larry. 1996. *Pulping the South: Industrial Tree Plantations and the World Paper Economy*. Zed Books Ltd. London and New Jersey.

Dictionary.com, LLC. Copyright © 2010. All rights reserved. <http://dictionary.reference.com/browse/justice> Oct. 11, 2010.

Encyclopædia Britannica. © 2010 Encyclopædia Britannica Online. 03 October 2010.

Molloy, Michael. 2002. Experiencing the World's Religions: Tradition, Challenge and Chance. California, USA.

Fundamentalism. (n.d.) *The American Heritage® Dictionary of the English Language, Fourth Edition*. (2003). Retrieved October 14, 2010 fromhttp://www.thefreedictionary.com/fundamentalism

Gain, Philip. 2006. *Stolen Forests.* Society for Environment and Human Development (SEHD), Dhaka.

GORDON MARSHALL. "positive discrimination." A Dictionary of

Merriam-Webster."racism." online dictionary. © 2010 Merriam-Webster, Incorporated. <http://www.merriam-webster.com/dictionary/> racism Oct. 12, 2010.

Harvey, David (2005): *A Brief History of Neoliberalism.* Oxford: Oxford University Press.

http://en.wikipedia.org/wiki/Animism

http://en.wikipedia.org/wiki/Shudra

http://en.wikipedia.org/wiki/Structural_adjustment

http://www.experiencefestival.com/sudra_caste

http://www.firstpeople.us/FP-Html-Wisdom/BlackElk.html

http://www.imf.org/external/np/exr/facts/conditio.htm

http://www.india9.com/i9show/Sudras-47540.htm

http://www.themystica.com/mystica/articles/a/animism.htm

http://www.thisisecocide.com/introduction/ecocide/

http://www.wisegeek.com/what-is-a-shudra.htm

http://www.wisegeek.com/what-is-a-structural-adjustment-program.htm

http://wwwnew.towson.edu/polsci/ppp/sp97/imf/POLSAP3.HTM

Indian Institute of Dalit Studies. 2008. Caste-based Discrimination in South Asia: A Study of Bangladesh. New Delhi, India.

Kohn, Margaret, "Colonialism", The Stanford Encyclopedia of Philosophy (Summer 2010 Edition), Edward N. Zalta (ed.), URL = <http://plato.stanford.edu/archives/sum2010/entries/colonialism/>.

Merriam-Webster Dictionary. © 2010 Merriam-Webster, Incorporated. 13 October 2010.

Neocolonialism. (n.d.). The American Heritage® New Dictionary of Cultural Literacy, Third Edition. Retrieved October 11, 2010, from Dictionary.com website: http://dictionary.reference.com/browse/neocolonialism

Princeton University "About WordNet." WordNet. Princeton University. 2010.http://wordnet.princeton.edu

Princeton University "About WordNet." WordNet. Princeton University. 2010. < http://wordnetweb.princeton.edu/perl/webwn=nationalism >

Oct. 11, 2010.

Roy, Raja Devasish. The ILO CONVENTION ON INDIGENOUS AND TRIBAL POPULATIONS, 1957 (No. 107) AND BANGLADESH: A COMPARATIVE REVIEW. International Labour Organization, 2009.

Sensu Cooper at al. 1992. in Woodwell, George M et al. 2001. FORESTS IN A FULL WORLD. Yale University, USA.

Sociology. 1998. Encyclopedia.com. 12 Oct. 2010<http://www. encyclopedia.com>.

Thekaekara, Mari Marcel. *Combating caste: the stink of untouchability and how those most affected are trying to remove it.* New Internationalist, July 2005.

UC Atlas of Global Inequality; Glossary. March 13, 2003. <http://ucatlas. ucsc.edu/glossary.html> Oct. 13, 2010.

http://en.wikipedia.org/wiki/Political_ecology. 16 May 2012.

Schlosberg, David. 2007. *Defining Environmental Justice: Theories, Movements, and Nature.* Oxford University Press.http://en.wikipedia.org/ wiki/Environmental_justice#cite_note-Schlosberg-1. 16 May 2012.

http://en.wikipedia.org/wiki/Internally_displaced_person. 16 May 2012.

http://www.unpo.org/article/13687. Accessed 4 March 2013.

Prepared by Philip Gain, Robert Alec Lindeman, Asif Khan, Ainud Sony

Contributors

Haroun Er Rashid: Professor of geography and dean, School of Environment, Independent University, Bangladesh.

Jenneke Arens: Has been involved in campaign work for the rights of indigenous peoples, in particular in the Chittagong Hill Tracts, since the 1980s.

Philip Gain: Director of SEHD, researcher, freelance journalist, and adjunct faculty of Dept. of Media and Communication, Independent University, Bangladesh.

Devasish Roy: Chakma Raja and Chief of the Chakma Circle in the Chittagong Hill Tracts, an advocate at the Supreme Court of Bangladesh and a member of the UN Permanent Forum on Indigenous Issues.

Raja Tridiv Roy (late): Former Chief of the Chakma Circle in the Chittagong Hill Tracts.

Sayam U. Chowdhury: Wildlife biologist and author of "A Pictorial Field Guide to Shorebirds of Bangladesh".

Buddhajyoti Chakma: Journalist.

Partha Shankar Saha: Former SEHD research staff.

Asfara Ahmed: Researcher on environment and development.

Sudibya Kanti Khisa: Freelance consultant, researcher, and writer particularly on livelihood and conservation issues in the Chittagong Hill Tracts.

Sarder Nasir Uddin, PhD: Senior scientific officer, Bangladesh National Herbarium.

M. Monirul H. Khan, PhD: Wildlife biologist engaged in teaching and research in the field of monitoring and conservation of wildlife. Currently he is an associate professor of zoology at Jahangirnagar University. He is also interested in wildlife photography.

Ronald Halder: Ornithologist and author of "A Photographic Guide to the Birds of Bangladesh" and "Guide to the Bird Songs of Bangladesh".

Han Han: Architect, Master's candidate in sustainability (Adv) at The University of Adelaide.

Ainud Sony: SEHD research staff.

A.K.M. Muajjam Hossain Russel: Partner, FORM.3 archtiects.

Tahmid Huq Easher: Lecturer and research associate.

Sadeka Halim: Professor of Sociology, Dhaka University and Information Commissioner, Information Commission, Bangladesh.

Shekhar Kanti Ray: Former SEHD staff.

Md. Safiullah Safi: Former SEHD staff.

Syeda Nusrat Haque: Intern with SEHD.

Supryio Chakma: Photo journalist.

Alimul Haque: Former SEHD staff.

Robert Alec Lindeman: Student, University of Massachusetts Amherst, USA.

Asif Khan: Student.

Tania Sultana: Documentation In-charge, SEHD.

Ushing Prue: Student.

Nimaprue Marma: Student.

Ching Mo Sang: Student.

Lucky Chakma: Student.

Acknowledgements

Innumerable people, communities, and organizations have assisted SEHD for more than a decade in the preparation of this report on the state of the environment in the Chittagong Hill Tracts (CHT). It is not possible to mention the names of all the individuals and organizations that have extended their generous support to SEHD staff and other contributors during their research and investigation.

First of all we gratefully thank all contributors. They patiently engaged with the SEHD staff and local communities in formal and informal meetings, discussions, seminars, workshops, etc. and have shared their knowledge and intelligence on the CHT environment.

The names of individuals that we must mention with gratitude include Goutom Dewan, Raja Devasish Roy, Ronald Halder, Dr. Monirul H. Sayam U. Chowdhury, Prof. Haroun Er Rashid, Sudibya Kanti Khisa, Supriyo Chakma, Harikishore Chakma, Dhung Cha Aung Chak, ZuamLian Amlai, Sudatta Bikash Tanchangya, Arun Laibrashaw, Ranglai Murung, Buddhajyoti Chakma, Arunendu Tripura, Annaya Shadhon Chakma, Hironmoy Chakma, Dibakar Dewan, Mong Mong Chak, Lelung Khumi, Ching Chala Chak, Cha May Chak, Mong Kew Ching Chak, Mong Chanu Chak, Sunil Kanti Dey, Noiring Khumi, Siku Khumi, and Prof. Ahaduzzaman Mohammad Ali among others.

Many of these individuals have contributed write-ups themselves on the one hand and on the other, have assisted us with guidance, photographs and information.

We are deeply indebted to Ainud Sony, Tania Sultana, Sabrina Miti Gain, and Josephine Gain who provided editorial and translation assistance, checked facts, and patiently proofread the texts after page design.

Brother Jarlath D'Souza has patiently gone through every page of this report for months; his language editing and editorial assistance have been helpful in improving the texts. We are deeply indebted and grateful to him. We are grateful to Prof. Raquib Ahmed for preparing some maps used in this book.

People at SEHD who have patiently assisted the editor, writers and contributors in the field and in logistical matters include Lucky Ruga, Prasad Sarker, Babul Boiragi, and S.N. Obaidul Muktader.

Misereor of Germany and ICCO of the Netherlands deserve special thanks for their support to SEHD and for making this production possible. They have always been generous and flexible concerning the choices and strategies of SEHD.

Abbreviations

ACF	Assistant Conservator of Forest
ADAB	Austrian Development Assistance Bureau
ADB	Asian Development Bank
ADC	Additional Deputy Commissioner
BAT	British American Tobacco
BFIDC	Bangladesh Forest Industries Development Corporation
BDR	Bangladesh Rifles
BGB	Bangladesh Border Guard
BMRE	Balancing, Modernization, Rehabilitation, and Expansion
BNP	Bangladesh Nationalist Party
CHT	Chittagong Hill Tracts
CHP	Community Hygiene Promoter
CHTDB	Chittagong Hill Tracts Development Board
CHTDF	Chittagong Hill Tracts Development Facility
CF	Conservator of Forest
DAE	Department of Agriculture Extension
DANIDA	Danish International Development Agency
DAT	Dhaka Tobacco
DC	Deputy Commissioner
DFO	Divisional Forest Officer
DGFI	Directorate General of Forces Intelligence
DoE	Department of Environment
EU	European Union
FAO	Food and Agriculture Organization of the United Nations
FD	Forest Department
FRI	Forest Research Institute
FSP	Forestry Sector Project
GBM	Ganges-Brahmaputra Meghna
GH	Green Hill

GoB	Government of Bangladesh
GOC	General Officer Commanding
ICS	Indian Civil Service
IDP	Internally Displaced Person
IFC	International Finance Corporation
IFI	International Financial Institution
ILO	International Labour Organization
IMF	International Monetary Fund
IP	Indigenous People
IUCN	International Union for Conservation of Nature
KPM	Karnaphuli Paper Mill
LGED	Local Government Engineering Department
MDBs	Multilateral development banks
MoCHTA	Ministry of the Chittagong Hill Tracts Affairs
MoEF	Ministry of Environment and Forest
MP	Member of Parliament
NEFA	North Eastern Frontier Agency
NGO	Non-Governmental Organization
PCJSS	Parbattya Chattagram Jana Samhati Samiti
PF	Private Forests
PF	Protected Forests
RC	Regional Council
RF	Reserved Forests
RWHS	Rainwater Harvesting System
SEHD	Society for Environment and Human Development
SIDA	Swedish International Development Authority
SKS	Sena Kalyan Sangstha
SOP	Sulfate of Potash
TANDP	Thana Afforestation and Nursery Development Project
TFAP	Tropical Forestry Action Plan
TNO	Thana Nirbahi Officer

TSP	Triple Super Phosphate
UN	United Nations
UNCED	United Nations Conference in Environment and Development
UNICEF	United Nations Children's Fund
UNDP	United Nations Development Program
UNPFII	UN Permanent Forum on Indigenous Issues
UPDF	United Peoples Democratic Front
UP	Union Parishad
USS	Upland Settlement Scheme
USF	Unclassed State Forests
VCFs	Village Common Forests
VDC	Village Development Committee
VDG	Village Development Group
VDP	Village Defense Party

Captions: Chapter Introductory Photos

Fresh rubber cultivation in steep hills in Alekhhyong *Mouza* in Naikhongchhari Upazila in Bandarban. The entire Alekhyong *Mouza* has been engulfed by rubber cultivation. It is mainly the outsiders who have lease of land for rubber cultivation. © Philip Gain

The last vegetation of the hills in Baishari Mouza has just (2012) been cut for preparation of rubber. It is Chak inhabited area. Commercial rubber cultivation in Baishari Mouza in Naikhhongchhari has taken much of the Chaks' traditional land. © Philip Gain

(from left) (a) A *jumia* woman is harvesting paddy from her jum in the lake area of Rangamati Hill District. (b) A forlorn Tanchangya woman in a forest village in Farua Reserved Forest. She represents the agony of life of the people allowed to live in the reserved forests. (c) Lone tree stands in mere undergrowth in a Reserved Forest. © Philip Gain

Sneho Kumar Chakma (67) is a boidaya (village doctor or herbal practioner) in Sadar union of Longodu. He shows the cut trunk of Dellutti, used to treat any sort of eye-pain. © Philip Gain

(from left) (a) Likri Hill, the core area of Sangu Reserved Forest. © Monirul Khan. (b) Pulpwood plantation in the Kaptai Pulpwood division at harvest time. © Philip Gain

(from left) (a) One of a dozen species of bamboo in the Chittagong Hill Tracts (CHT). © Philip Gain. (b) Water filled passage through a cave © Ronald Halder. (c) A Chak woman of Baishari harvests herbal plants for herbal dishes on the way to Badurjhiri, another forest village. The forest in the area has disappeared. But the Chaks still find some edible plants from the remnants. © Philip Gain

(from left) (a) Asian Elephant is the flagship species of the Chittagong Hill Tracts. © Monirul Khan. (b) Great Hornbill. Once found throughout the Chittagong Hill Tracts, is now restricted only to a few remote forest patches. It may soon disappear completely. © Ronald Halder. (c) A Brick kiln in the Chittagong Hill Tracts. © Philip Gain

Tobacco now surrounds Badurjhiri Chak Village. The invasion of tobacco around Badurjhiri started in 2009. Twelve out of 15 Chak families of this village got engaged in its cultivation in 2010. Badurjhiri village seen through tobacco on their precious land. © Philip Gain

An outsider taking away the last logs out of a remote area inhabited by the Chaks. This is an everyday scene and there seems to be no one there to stop the pillage. © Philip Gain

(from left) (a) A Chakma woman in a remote forest village fetches fountain water that comes through a pipe at a hillside collection point. (b) A beautiful house on plinth on the hillside. (c) Vegetable and fruit market in Bandarban town where jumia women bring their produces from the *jum*. © Philip Gain

A temporary military camp on a mountain, shaved, possibly to keep it safe from mosquitos. © Philip Gain

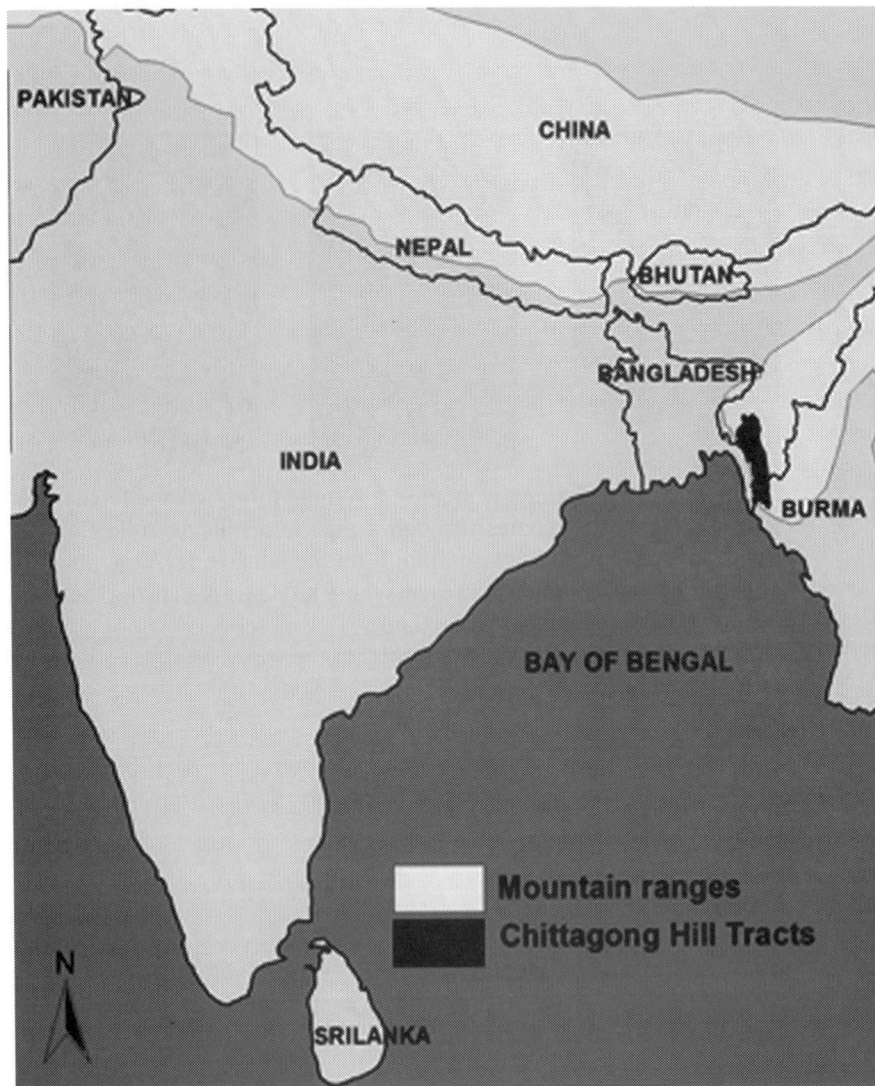